On the Trail of
Monkeys and Apes

On the Trail of Monkeys and Apes

Letitia FARRIS-TOUSSAINT and Bernard DE WETTER

Illustrations by Maël DEWYNTER

Contents

7	Introduction
9	**INTRODUCING MONKEYS AND APES**
10	Evolution and Characteristics
13	Classification and Distribution of Primates
16	Lifestyle
22	Social Life
29	**OBSERVING PRIMATES IN AFRICA AND IN THE MEDITERRANEAN**
30	Western Black-and-White Colobus
32	Ghana: Boabeng-Fiema Sanctuary
34	Mountain Gorilla
36	Congo, Rwanda, and Uganda: The Virungas Chain and Bwindi Forest
38	Barbary Macaque
40	Gibraltar: Monkey Hill
43	Chimpanzee
45	Tanzania: Gombe Stream National Park
49	Gelada
51	Ethiopia: Simian Mountains National Park
55	**OBSERVING LEMURS IN MADAGASCAR**
56	Madagascar, the Island Continent
59	Indri
61	Verreux's Sifaka
63	Ring-tailed Lemur
64	Lesser Bamboo Lemur
67	**OBSERVING MONKEYS IN ASIA**
68	White-handed or Common Gibbon
70	Thailand: Khao Yai and Khao Phra Thaew National Parks
74	Japanese Macaque
76	Japan (Honshu): Jigokudani Onsen
79	Celebes Black Ape
81	Indonesia (Sulawesi): Tangkoko Duasudara Natural Reserve
84	Orangutan
86	Indonesia (Sumatra): Bohorok Rehabilitation Center, Gunung Leuser National Park
90	Hanuman Langur
92	India (Rajasthan): Ranthambhore National Park

97	**OBSERVING MONKEYS IN AMERICA**
98	Black Howler
100	Belize: Community Baboon Sanctuary of Bermudian Landing
104	White-faced Capuchin
106	Costa Rica: Santa Rosa National Park
110	Red-backed Squirrel Monkey
112	Costa Rica: Manuel Antonio National Park
117	**MONKEYS, APES, AND MAN**
118	Respectful Tourism
119	A Death Sentence?
121	Helping Primates
123	**APPENDICES**
124	Equipment
124	Photographing Primates
124	Organizations
125	Glossary
125	To Learn More
126	Primate Index
127	Geographical Index

Introduction

Warning

The world changes quickly, as do travel conditions. The information that we have given corresponds to that which existed at the time the text was edited. Sociopolitical conditions may have changed considerably by the time you read this. Who would have thought only a few years ago that the Volcano National Park, the sanctuary for mountain gorillas in Rwanda, would be transformed one day into a hideout for armed rebels? Conversely, only a short time ago, who would have had the nerve to have ventured into an Ethiopia ravaged by war to spend one's vacation there?

Finally, the sites presented in this work almost all concern countries in which the infrastructure of roads is very small in comparison with the so-called Western countries. That is why we have not thought it a good idea, in general, to indicate distances in miles; in these cases, we have given an estimate of the time necessary to cover these distances.

The dozens of species of monkeys that live on the planet are distributed over vast areas, which does not make it any easier to observe them in the wild. Some are certainly easy to see, even too easy, according to those who have already camped in the national parks of Africa and who had to suffer the plundering by baboons and vervets. These attacks were a real plague for the campers, as these monkeys have lost their natural fear of *Homo sapiens*, but these few and very specific species are an exception to the general rule that monkeys are unobtrusive and timid animals. The majority of them live in forests, an environment that is dense and closed by definition, where the possibilities for observation often remain uncertain. In addition, most of them lead a basically arboreal life, descending only rarely—sometimes never—to the ground, living in trees at a height of 132–197 ft (40–60 m). Even the largest species are not necessarily easy to see—orangutans almost never leave the highest trees, chimpanzees are experts in the art of concealment, and gorillas generally flee as soon as they detect the presence of humans.

The amount of time that can be spent finding monkeys helps to determine the quality of observations, but only a privileged few can remain in one place for weeks, even months, in order to learn more about the habits of the monkeys that live there and to gradually accustom them to the presence of humans.

We have, therefore, selected places where the conditions have been "prepared," in other words, where the primates are used to humans, or places in which structures for individual visits have been developed. These places offer reasonable guarantees for observing monkeys under optimum conditions, but it must always be kept in mind that these are wild animals, and finding them can never be formally guaranteed.

INTRODUCING MONKEYS AND APES

Introducing Monkeys and Apes

Evolution and Characteristics

Evolution
What remains of what would actually be the oldest primate fossil bones? A single molar, found in Purgatory, Alaska. Its dating places it in the upper Cretaceous period, approximately 66 million years ago, toward the end of the age of the dinosaurs. The animal to which it belonged has the name *Purgatorius ceratops*, so called after its contemporary, the triceratops dinosaur.

The order of primates was thus one of the first to diverge from primitive mammals, perhaps beginning with an insectivore in the tropical forest that may have resembled a tree shrew (order of Tupaiiforms). Today, there are 233 known species of primates, which still share a certain number of traits inherited from this common ancestor.

In certain ways, the very supple behavior and general morphology of primates may have evolved since the first mammals. Today, however, few criteria separate them from other mammals, aside from the structure of the inner ear, the perforations at the base of the skull, or the more rectangular shape of the molars, with their rounded cuspids.

The specialization of primates rests paradoxically on their lack of specialization. While very highly specialized beings are no longer capable of adapting to a changing environment, primates have, since the beginning, opted for flexibility, both behavioral and physical.

The ancestors of modern primates established themselves in the primitive forest at a time when angiosperms evolved and became more numerous, thus creating a new ecological niche, in which flowers, fruits, seeds, and buds abounded; the adaptations that characterize primates are thus linked to this arboreal life.

Characteristics
The particular characteristics of the primate **hand** are of the greatest importance for its life in the trees. Its high level of dexterity makes it possible for the animal to climb with precision, and guarantees the successful seizing of fruit and prey. The same is true of fingertips (and the end of the tail in certain New World monkeys), with pads that contain large concentrations of Meissner corpuscles, which are nerve endings or organs of touch. To protect these sensitive hands, in place of claws, primates have fingernails of various shapes—narrow and curved in New World monkeys, wide and flat in the great apes and humans. The sweat of the hands enables them to get a better hold on the often slippery surfaces of branches.

Thanks to the fleshy cushions on its fingers, tarsiers (shown here *Tarsium spectrum*) can attach themselves to the smooth surfaces of vertical poles.

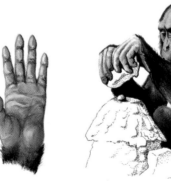

In Guinea, chimpanzees *(Pan paniscus)*, with the help of their agile hands, use stones to crack nuts.

Introducing Monkeys and Apes

The great agility of primates also depends on their **stereoscopic vision**, which gives them the ability to judge distance, required both for moving through trees and for capturing prey. The color vision of diurnal primates helps them to collect ripe fruit, flowers, and so on. These multiple advances in the evolution of touch and sight occurred at the expense of the senses of smell and hearing; thus, while the eyes moved toward the front of the face to provide vision in relief, the muzzle became shorter, considerably reducing the number of nerves in the nose and the size of the olfactory bulbs in the brain. In addition, mobile ears with thin walls were replaced by small, more or less immobile ears.

Modifications of the skull, the result, among others, of the reduction in the muzzle, made room for the **brain**, which had grown considerably in comparison with the body size. Unable to grow in an unlimited way, the brain folded over itself in convolutions, increasing the surface allocated for gray matter and, consequently, the ability to study the environment. In this way, learned behavior became more and more important in primates. In order to benefit, in total security, from a long period of apprenticeship, young primates remain dependent on their mothers for a much longer period of time than the majority of other mammals.

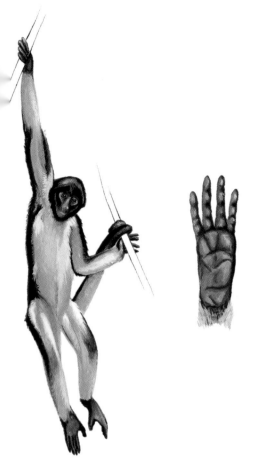

In the long-haired spider monkey *(Arteles belzebuth)*, the absence of a thumb facilitates brachiation.

Tonkean macaques *(Macaca tonkeana)* have a short thumb, opposable, adapted for moving with the hand flat on the ground.

Introducing Monkeys and Apes

Verreaux's sifaka
(*Propithecus verreauxi*)

Golden lion tamarin
(*Leontopithecus rosalia*)

Orangutan
(*Pongo pygmaeus*)

Mandrill
(*Mandrillus sphinx*)

Owl-faced guenon
(*Cercopithecus hamlyni*)

Celebes black macaque
(*Macaca nigra*)

Primates demonstrate great specific diversity, both on the level of features as well as on the level of morphology.

Introducing Monkeys and Apes

Classification and Distribution of Primates

According to the paleontologist Elwyn Simons, there is no "correct" classification of primates; in fact, the taxonomy of species evolved much more quickly than the species it describes. Here, the order of primates will be divided into two suborders: on one hand, the prosimians, which have kept a large number of the characteristics of their ancestors, on the other, the simians (or anthropoids), which, for 30 million years, have diverted from the primitive forms, acquiring a larger brain and greater manual dexterity. The simians divided into two subgroups during the upper Eocene, when some left Africa and arrived in America. The first became the Platyrrhini, the others the Catarrhini.

A General Survey of Prosimians

Prosimians are primates that are considered to be "primitive," to the extent that they have retained more ancestral traits than the

In Guyana, the Midas tamarin (*Saguinis midas*) populates the forests and even some residential areas. This small callitrichida lives primarily in South America.

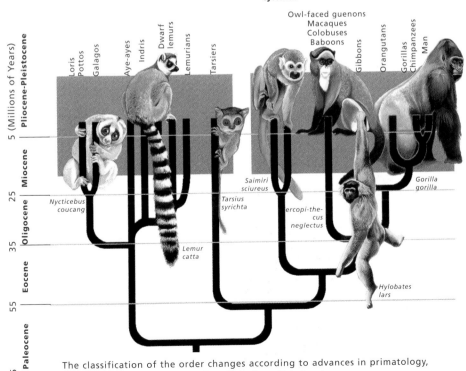

The classification of the order changes according to advances in primatology, but several large groups remain relatively well defined.

Introducing Monkeys and Apes

simians, such as a longer muzzle and a good sense of smell, binocular vision having only been able to develop partially. Some of the fingers end in claws rather than real nails, and, while the number of teeth varies, all prosimians have specialized teeth to be used as a comb in grooming.

Prosimians are generally considered to fall into three suborders belonging to two groups: the **tarsiforms**, including tarsiers, a nocturnal species limited to the Philippines, Borneo, Sumatra, and Sulawesi, and the **lemuriforms**, including, on one hand, the galagonidae (galagos) and loridae (pottos and loris), and, on the other hand, the lemuridae. Galagos and pottos, both nocturnal, are limited to Africa; the loris, also nocturnal, lives in Asia. Lemurs, formerly distributed throughout the rest of the world, now survive only in Madagascar and the Comoro Islands where there were never any monkeys.

The Simians

The simians include monkeys—of which there are approximately 200 species—and man.

The **New World monkeys**, or **platyrrhinian monkeys**, probably arrived on this continent by means of natural rafts, at a time when South America was closer to Africa than to North America. They can, in general, be identified by their nostrils, which open sideward, and not toward the bottom like those of their Old World ancestors. Their brain is relatively smaller and their dentition more primitive. The hallux is as powerful and apposable as in the catarrhinians, but the thumb is less so. All New World monkeys live in the tropical forest, and none have adapted to life on the ground.

The suborder combines two families; that of the **callitrichidae** includes the three types of marmosets and tamarins, the smallest of the real monkeys. The callitrichidae have claws on all of their digits, except for the hallux, which has a nail. They live in family groups, and twins, which are the rule, are carried by the father. The species of the family **cebidae** more closely resemble what we generally consider the monkey type. All have real nails on their fingers and toes. Species with a prehensile tail exist only in this family.

Catarrhinian monkeys include **Old World monkeys**, gibbons, great apes, and man. Their larger, more complex brain permits better eye-hand coordination. Certain species have adapted to a life that is partially, and sometimes completely, on the ground. The infra-order is broken down into two super-families.

• The superfamily of the **cercopithecoids** includes only one family, that of the cercopithecidae. These are the most numerous, the most varied, and the most widely distributed (with the exception of man) of all of the primates, including, in Asia and Africa, approximately 17 genera and 85 species. The majority have ischial callosities of hard skin, serving as natural cushions in the sitting position. The subfamily of **colobins** is primarily Asiatic. Their stomach is adapted to a diet of leaves. The subfamily **cercopithecines** is distributed between Africa and Asia, where habitats and diet are very varied.

The superfamily of **hominoids** groups together gibbons, the great apes, and man. All have common traits inherited from their tree-living ancestors: the vertical position, very supple joints in the arms and shoulders, and, with the exception of man, arms that are longer than the legs. These are the largest of the primates. There is a long period of gestation and the young remain with their mothers for a particularly long time. Other

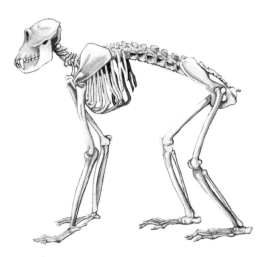

Old World monkeys are often well adapted to life on the ground. The arms of terrestrial primates (shown here, the skeleton of a macaque) are relatively much longer than those of their arboreal relatives, with the exception of monkeys, which move by brachiation.

common characteristics are the absence of a tail, large, complex brains, and even more accurate vision.
• The family **hylobatidae**, limited to Asia, consists of gibbons and simiangs. Smaller than the great apes, they have ischial callosities like the cercopithecidae. The only true brachiators, they live almost exclusively in the trees. The **pongids** and the **hominids** are distinguished both by their large size and their high level of intelligence, as well as their habit of sleeping in a nest or a shelter. The **pongids**, originally from Asia, include only one species, the orangutan. Its anatomy is particularly adapted for moving in trees. The **hominids**, originally from Africa, have two subdivisions known as tribes. The **paninian** tribe includes gorillas and the two species of chimpanzees. These are distinguished by their quadruped locomotion leaning on the front of the external face of the phalanx. The **hominian** tribe contains a single species, man. The latter is distinguishable by its biped locomotion and the anatomical adaptations associated with it, as well as by its brain, which has a volume two times larger than that of the great apes, and the exceptional technological, cultural, and linguistic complexity of human societies.

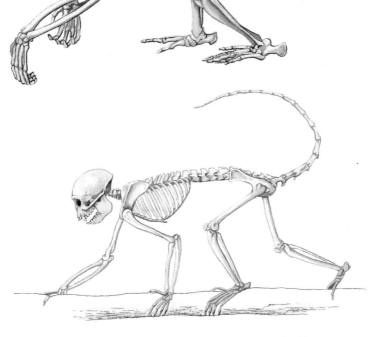

Skeletons of a gorilla *(Gorilla gorilla)* and a black-bearded saki *(Chiropotes satanas)*, respectively adapted to live on land and in the trees.

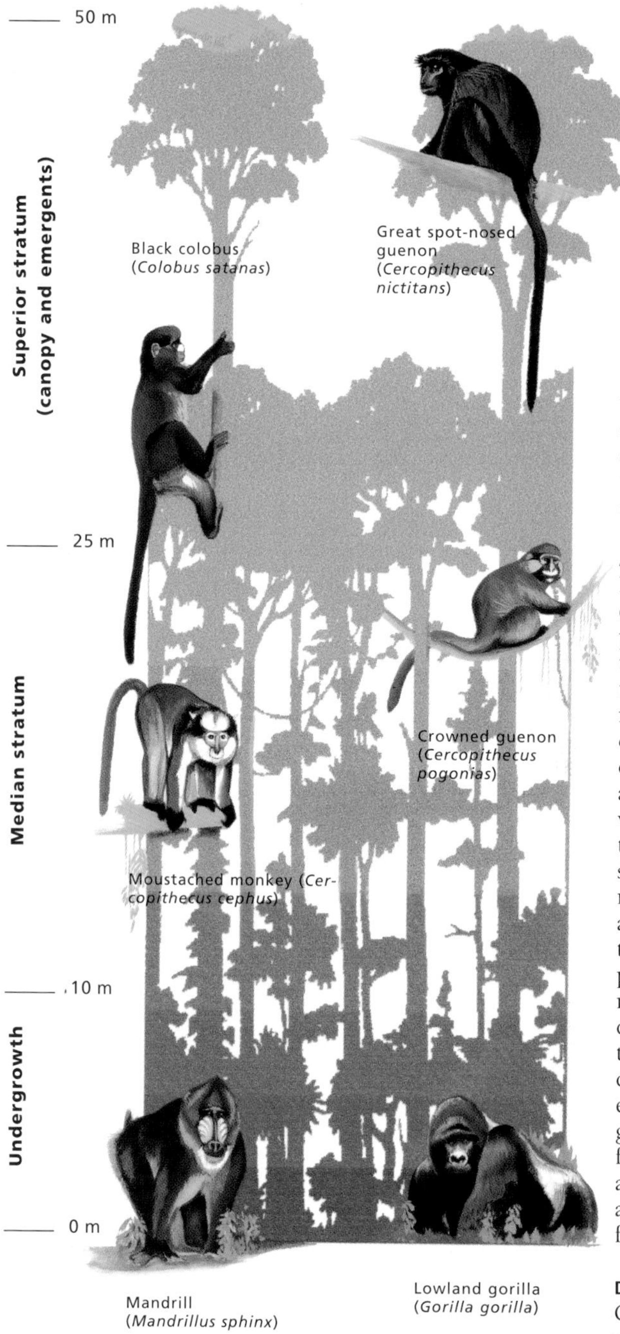

Introducing Monkeys and Apes

Superior stratum (canopy and emergents)

Median stratum

Undergrowth

- 50 m
- 25 m
- 10 m
- 0 m

Black colobus (*Colobus satanas*)

Great spot-nosed guenon (*Cercopithecus nictitans*)

Crowned guenon (*Cercopithecus pogonias*)

Moustached monkey (*Cercopithecus cephus*)

Mandrill (*Mandrillus sphinx*)

Lowland gorilla (*Gorilla gorilla*)

The sharing of space and the use of various forest strata by different primates in a tropical forest in Gabon, Africa (monkeys are of course not on the scale).

Lifestyle

Life in the Forest, Life on the Savannah

Eighty percent of primates live in the tropical forest, where they occupy all levels. Their presence there is dominant. More than half of the species, however, are capable of adapting to life in the dry forest. Certain Old World monkeys have thus succeeded in settling in the savannah and semidesert areas, but to sleep, all take refuge in the trees or the rocky cliffs, far from predators.

Antipredator Behavior

Monkeys have many ways of defending themselves against predators. Some are content to flee or to camouflage themselves, others create noisy disturbances to discourage the enemy, others come together to attack it in a group, while the baboons, with well-developed canine teeth, will brave the aggressor alone, if the situation so requires. All emit cries of alarm to warn the rest of the troop of the presence of a predator. Often, the type of menace is indicated in the call itself, which provokes the appropriate response; a certain cry signals the presence of a panther on the ground, and the monkeys flee toward the treetops; another announces an eagle, and they dive into the foliage, out of harm's way.

Diet

Generally omnivorous, primates consume fruits and leaves along with insects, larvae, eggs, small vertebrates, and various tree gums and saps. The size of the animal influences its type of diet.

INTRODUCING MONKEYS AND APES

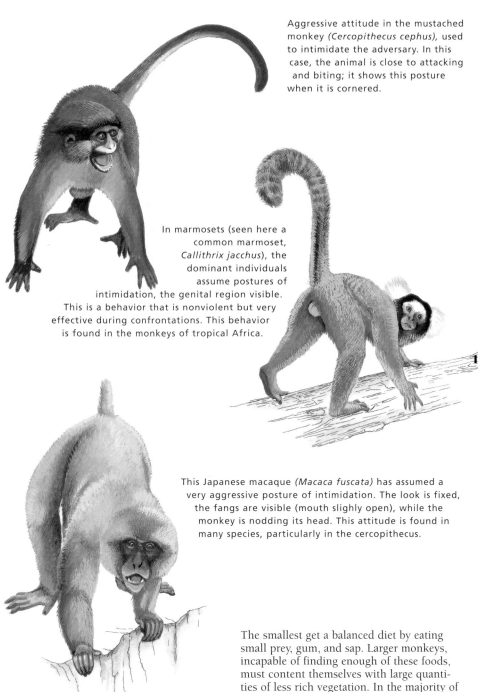

Aggressive attitude in the mustached monkey *(Cercopithecus cephus)*, used to intimidate the adversary. In this case, the animal is close to attacking and biting; it shows this posture when it is cornered.

In marmosets (seen here a common marmoset, *Callithrix jacchus*), the dominant individuals assume postures of intimidation, the genital region visible. This is a behavior that is nonviolent but very effective during confrontations. This behavior is found in the monkeys of tropical Africa.

This Japanese macaque *(Macaca fuscata)* has assumed a very aggressive posture of intimidation. The look is fixed, the fangs are visible (mouth slightly open), while the monkey is nodding its head. This attitude is found in many species, particularly in the cercopithecus.

Intimidation and aggression are expressed differently according to the species.

The smallest get a balanced diet by eating small prey, gum, and sap. Larger monkeys, incapable of finding enough of these foods, must content themselves with large quantities of less rich vegetation. In the majority of species, fruits constitute an important part of the diet, with the exception of geladas, the only grazing monkeys, and the colobines, which feed almost exclusively on leaves.

Introducing Monkeys and Apes

The digestive system of the colobus permits it to consume leaves almost exclusively.

The leopard defends its prey against the baboon, which has a well-developed taste for meat.

INTRODUCING MONKEYS AND APES

The omnivorous diet of the macaque includes some very sweet fruits.

Golden lion tamarins, *Leontopithecus rosalia,* follow a frugivorous diet with a strong percentage of insects.

Introducing Monkeys and Apes

On the ground, quadruped locomotion occurs with the hands either flat, as in the macaque, or on the phalanges, as in the chimpanzee.

Locomotion

The classification of types of locomotion in primates is as controversial as the taxonomic classification. There are generally four principal types, with multiple variations.

In **vertical locomotion**, the animal attaches itself vertically to the trunks of trees and moves by jumping this way from one tree to another. This is the most common form of locomotion among the prosimians, but in monkeys, it is only practiced by marmosets and tamarins.

In **quadruped locomotion**, the animal moves on four feet, its weight equally distributed on each. In the trees, it grips the horizontal branches with its hands and its prehensile feet. On the ground, it walks with flat feet, leaning forward on the digits. Chimpanzees, gorillas, and, more rarely, orangutans lean forward on the external face of the second knuckle.

In **brachiation**, the animal, suspended by two hands, moves by swinging from branch to branch. While many monkeys occasionally practice this form of locomotion, only gibbons and siamangs are true brachiators.

Introducing Monkeys and Apes

Gibbons are the only true brachiators.

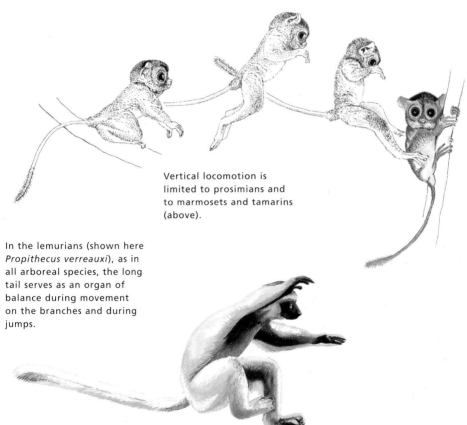

Vertical locomotion is limited to prosimians and to marmosets and tamarins (above).

In the lemurians (shown here *Propithecus verreauxi*), as in all arboreal species, the long tail serves as an organ of balance during movement on the branches and during jumps.

Introducing Monkeys and Apes

The prehensile tail (present only in certain New World monkeys) can be used as a third hand. Shown here, a red howler (*Alouatta seniculus*).

The hamadryas baboon (*Papio hamadryas hamadryas*). Unlike other baboons, they can cuddle up next to each other during periods of rest.

In **biped locomotion**, the animal moves with all of its weight distributed on its two rear feet. Only man possesses an anatomy that makes it possible for him to move comfortably in this manner.

Social Life

Monkeys are social animals. The fossilized bones of the first primate already show a sexual dimorphism, which suggests that since their beginnings they have lived in groups. In groups where the competition between males is rough and constant, natural selection has benefitted those that are sufficiently large and strong to win the right to mate. This social life offers many advantages: First, it provides for the safety of each, as the more there are, the less danger there is; then, it guarantees the proximity of sexual partners; finally, the experience of the eldest serves as an example to the young. The various aspects of the social life of monkeys described here varies considerably according to the species.

Composition of Troops

Unimale troops consist of several families and their young, and a single male that does not tolerate any rivals. A young male, pushed out at puberty by his father, can, while waiting for the opportunity to form his own "harem," remain alone or join a group with other single males. The harem is created either by carrying off very young females and keeping them until maturity, or by seducing the females of other troops, or, finally, by deposing the dominant male of an existing troop.

Species forming **multimale troops** generally have a complex social organization, including, among other things, privileged male-female relationships, the transmission of social rank from the mother to her young, and a hierarchy among males that may be complicated by the formation of friendships. The relationships in multimale troops of Old World monkeys are generally more contentious than among New World monkeys.

The **family group** is a relatively rare form, but is present in all of the major branches of primates (prosimians, Old and New World monkeys, and hominoids), which indicates that it is not phylogenetic. These groups, formed by an adult couple accompanied by a maximum of four offspring, habitually

Introducing Monkeys and Apes

occupy a well-defended territory. These species show little or no sexual dimorphism.

There are **several exceptions**. The orangutan, for example, leads a relatively solitary life; the male has a large territory that includes the small territories of several females. Their meetings, of variable length, remain limited to the mating period. Among the brown-headed tamarins of Peru, there exist troops with a single female for several males. As of this date, this is the only case of polyandry known among primates: Geladas form unimale bands within troops of several hundred animals.

Communication

The complexity of the social life of monkeys implies an efficient system of communication. Transmitted from generation to generation, this system can be extremely elaborate. Among diurnal primates, **visual communication** dominates. The entire body may be involved. Posture, position of the tail, hair raised or not, and facial expression are very important. Some signals are very subtle; for example, many species lower their eyelids, which are white, in order to communicate. But this gesture can just as easily be used by a dominant individual to threaten a subordinate as by a subordinate to appease a dominant individual. Likewise, the "smile," common in most evolved simians, can serve as an invitation to play, but is also found in an exaggerated and strained form when an animal finds itself in the presence of its hierarchical superiors. Everything is a question of nuance.

In the forest, where gesticulations and postures are more difficult to see through the foliage, **auditory communication** is more important. It is, therefore, not surprising that it is more varied among the arboreal monkeys. Some sounds are similar to our vowels and consonants; others are more difficult, even impossible, to detect. Vocalizations indicate the age, sex, and social rank of individuals. They make it possible for the group to come together, to get rid of intruders, and to signal the presence of predators. Grooming among monkeys constitutes a basic form of **tactile communication**. In general, when

Among hamadryas baboons, the family harem is the basic social unit. During travel, they always walk in single file, the male in the lead, followed by the females (between 1 and 20), and the juveniles.

Facial attitudes of chimpanzees and their meaning:
A. Play (mouth open, upper teeth covered)
B. Fear or intense excitation (mouth open wide, fangs visible)
C. Mild excitation (lips forward)
D. Sullen expression (attitude observed in case of attack; the hairs on the face are standing on end)
E. Submission (when faced with a fellow creature of higher social rank)

asking to be groomed, an animal shows a part of its body to another. The latter responds by examining its fur and its skin in order to remove dirt and parasites, then, the positions are reversed. In addition to the obvious pleasure that monkeys can find in it, grooming also dissipates the tensions in a group, comforting an upset animal, appeasing an aggressor, or making it possible for a lower-ranked animal to create amicable relationships with a dominant one. There are other forms of tactile communication, such as the kiss or the intertwining of tails among friends, and the act of mounting a hierarchically inferior individual from behind to remind it of its rank.

More important among prosimians, which have, in part, retained the olfactory system of their ancestors, **olfactory communication** is nevertheless present in all simians, including man. The primates are among the rare mammals to have sudoriferous glands and some species have odoriferous glands exclusively designed for this type of communication. As

Introducing Monkeys and Apes

Scene of grooming among baboons. Grooming in monkeys is an important phase of tactile communication among the members of the group.

with auditory signals, olfactory signals serve to transmit information relative to the age, sex, social rank, and emotional state of individuals, as well as to define the territory of troops.

Dominance and Hierarchy

To determine the hierarchical order of two animals, one needs only to put a piece of food between them. The individual that takes the food first is the dominant one; the one that watches is the subordinate. This follows for the rank of each individual. Hierarchy is present among all primates, in various degrees and in various and coexisting forms.

The hierarchical order **among males** will be determined by the outcome of various confrontations and fights. Occasionally, males of the second and third rank will unite their efforts in order to usurp the dominant male. In this case, the co-vanquishers will then fight for first place.

Social rank **among females** extends through the maternal line, and is transmitted

Hierarchical order is an integral part of the social scene of Japanese macaques.

Introducing Monkeys and Apes

Young baboons generally adopt the dorsal position around 6–12 weeks, but some never do.

from generation to generation. It is independent from that of males. Unlike the latter, which often changes, the hierarchical order of females is extremely stable. In fact, despite the efforts of some ambitious females that try to raise their rank through friendships with dominant females, even by encouraging their offspring to play with the offspring of the latter, it is exceptional for them to succeed.

For the large majority of primates, it is the male that dominates, but the relationships **between the sexes** are complex. For example, for a long time it was believed that, because the dominant male is the first to mate with females in heat, he has more offspring than his subordinates. It is now known that in many species, females are not fertile until the end of heat, at the moment when they go off with their "friends," males with which they have amicable relationships in daily life. Low-ranked females, in particular, take advantage of these relationships, which often make it possible for them to have access to better sources of food.

Reproduction
Reproduction varies considerably according to the species. In general, the cycle in primates is rather long, with the young remaining with their mothers for a long time and reaching sexual maturity late, while the adults live to a relatively old age.

The majority of monkeys have a period of estrus, or **heat**, which coincides with ovulation. Females of Old World species generally present an edema and/or reddening of the skin in the perineal area, called the sexual skin, occasionally accompanied by bleeding or specific odors. While human females do not show such signs of heat, various studies have shown that when they deceive their partners, deception occurs most often, without their being conscious of it, at the moment of ovulation.

For most monkeys, **mating** is limited to the moment of ovulation, and is often solicited by the female. A female olive baboon forms long-term friendships with one or two males who protect her and her offspring from aggression by other troop members. The female tends to mate with one of these friends the next time she is in heat. Normally, the male mounts the female from behind; however, gibbons and orangutans can mate face to face, and even hanging from branches. The bonobo (pygmy chimpanzee) is the only primate, outside of man, to experiment with a wider variety of positions, but it is man who devotes the most time to this activity.

The gestation period considerably exceeds that of the majority of mammals of comparative size, reaching nine months for the gorilla (258 days), the orangutan (264 days), and man (256 days). Few signs indicate the approach of **birth**. In monkeys, this takes place at night and does not last more than two hours. As for the great apes, which have less chance of being attacked by predators, giving birth may take place at any time. A single offspring is the rule, except among certain prosimians and New World marmosets and tamarins. No newborn is capable of following its mother at birth, but the majority are able to attach themselves solidly to her fur to be transported, generally under the belly and, later, on the back. Babies nurse from two teats located on the chest. They develop rapidly, but remain dependent on their mother or their family for a long time, even after weaning, as apprenticeship takes time. There is, moreover, a correlation between the length of **childhood** and the size of the brain.

In the majority of species, young males reaching **sexual maturity** leave their birth troop. This involves numerous risks and explains why the number of females is so often higher than that of males. These young males can depart on their own (as, for example, among cynocephalous baboons), or be chased away by the dominant male, as, for example, among hanuman langurs. It also happens that it is the females that migrate, for example, among gorillas. In family groups, the male generally chases away the first-born male, and the female the first-born female (among the indris).

The extended childhood of the bonobo is closely linked to the size of the brain, and thus to its high level of intelligence.

OBSERVING PRIMATES IN AFRICA AND IN THE MEDITERRANEAN

Western Black-and-White Colobus

(Colobus polykomos)

FRENCH: Colobe blanc et noir d'Afrique occidentale
GERMAN: Barenstummelaffe
SPANISH: colob blanco y negro de Africa occidental
LOCAL: mbega, kuluzu

Description

Average length: male body plus head 23 inches (59 cm), tail 32 inches (81 cm); female body plus head 22 inches (55 cm), tail 30 inches (77 cm).
Average weight: male between 19 and 32 pounds (9 and 14.5 kg); female between 14 and 22 pounds (6.5 and 10 kg).

The number of subspecies varies enormously. All have a nose that hangs slightly over the mouth and a reduced thumb. Most have a totally white tail, but one of the many subspecies is entirely black.

Family

Cercopithecidal.

Locomotion

Black-and-white colobus rarely descend from the trees, where they move, whatever their speed, in a manner similar to that of the squirrel. Capable of prodigious leaps, it lands feet first, before jumping again.

Diet

All colobus have a large, complex stomach, similar to that of ruminants. This makes it possible for them to eat every day—when young shoots, flowers, and fruits are rare—up to 4 to 6 pounds (2 to 3 kg) of leaves that are too fibrous or toxic for other monkeys, and sometimes from a single species of tree.

Predators

Primarily eagles, which male colobus can attack when they land nearby. To avoid them, other members of the troop dive for distances of 49 feet (15 m) into the undergrowth.

Longevity

In captivity, more than 23 years.

Distribution

The subspecies *Colobus polykomos polykomos* is distributed from Gambia to the Ivory Coast; they are found in primary and secondary forest, forest islands in the humid savannah, and in the gallery forests along rivers and swamps.

Status

On the whole, this colobus is still relatively abundant, but some species are very much endangered.

Social Organization

Troops normally consist of six to ten individuals including a single mature male, which defends a well-defined small territory, and a matron that determines travel. The abundance of food that the folivorous diet provides makes it possible for a large number of animals to maintain good relations; thus, the colobus assigns a great deal of importance to grooming. An individual asking another for grooming can slap the face of the other that pretends not to see, while another wanting to groom a companion can grab it by the hair, and immobilize it in order to impose its ser-

Western Black-and-White Colobus

vices. Moreover, females and young males often grab the very young, still white, to pamper them. After three weeks, the youngster is better capable of withstanding kidnapping, by crying and kicking, then, when its coat gets darker, at around 14 to 17 weeks, it becomes less attractive.

after five weeks. They begin to eat solid food between three and six months.

History

In the Omo Valley in Ethiopia, the discovery of a fossil jawbone dating back two million years suggests that the current genus of colobus may have descended from an animal the size of a large baboon.

Behavior

Due to its diet and its digestive system that is similar to that of a ruminant, the colobus is not very active; even when feeding, it remains seated 99 percent of the time. Arising late, it heads toward the top of the trees for its first sunbath, visits its neighbors, and grooms itself for approximately one hour. Then the troop proceeds in a leisurely manner toward the trees, where it feeds actively until the midday heat obliges it to nap. When it becomes cool, the colobus eats again, before returning to the dormitory.

Reproduction

After six months of gestation, the female gives birth to a completely white infant with a pink face, of a good size—16 ounces (450 g)—and well developed. Births are not rigorously seasonal, but are more common during the rainy season. The infants are carried on the mother's belly until the age of eight months, but they move alone

In one day, the black-and-white colobus consumes one third of its weight in leaves and fruit.

Observation Site

Ghana: Boabeng-Fiema Sanctuary

Animals are an important element in the culture and the daily life of the Ghanaian people. Thus, while the pronounced taste of the people for meat from the bush often has rather unfortunate consequences for the fauna, ancestral beliefs effectively contribute to the protection of certain species.

Numerous tribes have animal totems that they are forbidden to kill, to hunt, to eat, or simply to disturb. These animals have an important place in most religious rituals and popular celebrations, and "sacred sanctuaries" have been established throughout the country for their benefit. These sacred sites are considered as refuges for the animal totems and protective spirits; often, they are also places where dead tribal chiefs are buried. Almost 2,000 of these sanctuaries exist in Ghana, including that of Baobeng-Fiema.

The latter extends along the edges of the two villages from which it gets its name. It is located approximately 62 miles (100 km) north of the city of Kumasi, former capital of the Ashanti dynasties, located in the Brong-Ahafo region in the center of the country. This sacred sanctuary, while only covering 4,502 acres (450 ha) in total, contains one of the densest concentrations of primates in the country. It is the only one for the moment to have been officially recognized by the government authorities, which gave it the status of national natural reserve in 1975.

According to village beliefs, the monkeys living in the sanctuary are the direct descendants of the male god Abodwo, who created the village of Fiema, and the female goddess Daworoh, who gave birth to the village of Boabeng. Legend has it that the ancestors of the inhabitants of these two bush villages one day found a colobus and a mona monkey, seated, wearing the robes of a tribal chief, and they decided to build the villages where the monkeys had indicated. A very strict taboo today prevents the cutting down of trees in the sacred forest, and the killing of monkeys, which have lived there since time immemorial in perfect harmony with the villagers.

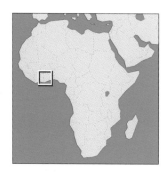

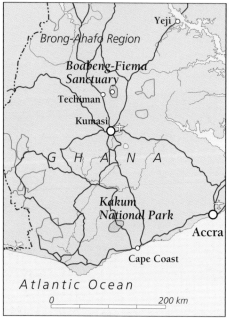

Fauna and Flora

The sacred forest sanctuary contains various types of vegetation: primary forest, secondary forest that grew following fires—primarily the result of noncontrolled fires in cultivated areas bordering the forest—and areas of bush savannah.

The divine status given to monkeys does not, unfortunately, apply to the other animals that populate the sanctuary, which explains why these have become rare in Boabeng-Fiema. Birds remain numerous in the forest, but large mammals have practically disappeared as a result of hunting by villagers for bush meat; only some duikers still survive there. As the human population increases and urbanization takes its toll, the pattern becomes sadly familiar.

Observation

The two species of primate that populate the sacred sanctuary are the black-and-white colobus and the mona, two species typical of the forest. In 1995 more than 120 colobus distributed in 8 troops, along with approximately 200 monas, distributed in 13 troops, frequented the sanctuary and the villages.

The possibilities for observing monkeys in the sacred sanctuary are excellent, and often surprising, as well. Where else in Western Africa would it be possible to see wild monkeys living in perfect harmony with villagers, evolving from generation to generation, without showing signs of fear of humans. The colobus are rather exuberant, while the monas exhibit a temperament that is somewhat more timid.

It is in fact easier to see and to observe the animals on the immediate outskirts of the villages than in the forest itself, where the closed environment and the arboreal habits of the animals make observation more difficult. The colobus often gather together in troops of several dozen individuals on the lower branches of the large trees, and frequently go as far as entering the villages, particularly to feed on the manioc peels that the residents leave out for them.

PRACTICAL INFORMATION

The best time for visits is from December to March, the period after the rains.

TRANSPORTATION

■ **BY PLANE.** Accra, the capital of Ghana, is linked to the United States by regular flights.

■ **BY BUSH TAXI.** Kumasi can be reached by local transportation or, individually, by interior road. Local transportation makes it possible to reach Boabeng and Fiema from Accra but requires long hours of travel on the road under minimum conditions of comfort and frequent changes and waits.

■ **BY RENTAL CAR.** The most realistic option for visiting the sanctuary consists of renting a vehicle in Accra, the capital, or possibly chartering a taxi for the round-trip. Making the round-trip and visiting the sacred sanctuary in a single day is, however, rather risky.

As for the Kakum national park, most travel agencies in Accra organize trips, but it can also be visited independently by renting a car with a driver who will know how to get there.

LODGING

There are not too many possibilities for lodging in the villages, but it is possible to obtain some space from one of the villagers—with minimal comfort, but a warm welcome and local color guaranteed. It is also possible to camp near the village, although this practice does not appear to be encouraged by the inhabitants. The only option, for those who are not attracted by the basic conditions of the accommodations, is to return to Kumasi, where they will find establishments that offer more comforts.

CLIMATE

The climate is tropical and rather humid. After the month of March, the harmattan blows from the north, making the air drier and relatively cooler. It also brings the dust from the sand of the Sahel, which, combined with the smoke from innumerable brush fires, makes the air hazy and reduces visibility.

TRAVEL CONDITIONS

At one time, villagers spontaneously welcomed visitors who had come to discover the sacred sanctuary and observe the primates. Custom required that permission first be requested from the tribal chief of one of the villages. Today, visitors are still warmly welcomed by the population, which is very hospitable, but it is no longer necessary to obtain the approval of the local authorities; on the other hand, there is now an entry fee, an inevitable consequence, perhaps, of the proportional growth in the number of visitors, which currently number about 1,000 per year. This admission fee remains rather low; the money received becomes part of a fund used to provide communal equipment to the villagers.

Mountain Gorilla
(Gorilla gorilla beringei)

GERMAN: Berggorilla
FRENCH: gorille de montagne
SWAHILI: makaku or gorila

Description

Standing height: male 55–73 in (140–185 cm); female up to 59 in (150 cm).
Average weight: male 352 lb (160 kg); female 187 lb (85 kg).

The gorilla is the largest of the primates. Its robust body is rounded; its muscles and its arms are short. Sexual dimorphism is marked. The head, massive in both sexes, ends in a conical shape among adult males. The latter is twice as heavy as the female and, as an adult, the color of its coat earns him the name of "silverback." Facial characteristics vary widely from one individual to another, with a strong family resemblance among members of the same maternal line.

Family

Pongidae.

Locomotion

The gorilla is basically a terrestrial animal. While the species is capable of climbing carefully among the trees, and even of building nests, the adult male does so only rarely. The gorilla moves on flat feet, but leans forward on the external face of the phalanxes. During the well-known displays when it beats its chest, the male can run 20 ft (6 m) on its two hind feet.

Diet

The mountain gorilla can consume some 58 species of plants, but three alone constitute 60 percent of its diet, basically consisting of leaves, shoots, and stems, with 2 percent fruit, and occasionally roots, bark, larvae, snails, and earth.

Predators

Gorillas that have already seen hunters at work can run away for miles upon a single viewing of man. On the other

For gorillas, early childhood lasts approximately three years, but it is still necessary to wait five years before the female reaches full maturity, and a dozen years for the young male.

hand, the male is generally ready to defend his harem to the death against leopards, but also against man, if the situation requires.

Longevity

The female lives from 40 to 50 years, the male 50 to 60.

Distribution

The mountain gorilla lives in the misty forests of the Virunga volcanic mountain chain, within the confines of the Democratic Republic of the Congo, Rwanda, and Uganda, at between 1.7 and 2.1 mi (2,800 and 3,400 m) of altitude, occasionally going as far as the Afro-alpine meadows, where there is hardly any food. Other gorillas, that is to say the remaining 98 percent, live in the rain forests of the lowlands.

Status

It is estimated that there are approximately 320 or 600 mountain gorillas, depending on whether or not the Bwindi population is included. The mountain gorilla is one of the rarest and most endangered mammals in the world. There is a glimmer of hope, however, despite the political problems in the former Zaire: Between the beginning of 1997 and the summer of 1998, ten births were recorded in the groups regularly tracked by scientists.

Social Organization

The mountain gorilla generally lives in troops of six to eleven members, capable of reaching up to 40 individuals. A typical troop includes a silverback, a younger male with a back that is still black (eight to twelve years), three adult females, and their offspring younger than eight years old. When a young female becomes an adult, she changes troop, becoming a member, where she reproduces for the first time. In general, she looks for a newly constituted troop, or even for a single male, because the social rank among females depends on the order of arrival. She remains as long as her male is still in charge. Most often, she changes troop again during her lifetime.

Outside of the mating season, parents have few relationships, except to energetically fight over the right to groom the offspring.

Behavior

The gorilla awakens at between 6:00 and 8:00 A.M., feeds until resting from 10:00 to 2:00 P.M., eats again until 5:30 P.M., then prepares its nest, generally on the ground. The construction of the nest, rather simple, seems a behavior inherited from the days when its ancestors spent more time in the trees. An animal of remarkable intelligence and dexterity, despite the crude appearance of its hands, the gorilla handles the most fragile objects—cameras, for example—with interest and care.

Reproduction

Reproduction among gorillas is slow, with four-year intervals between births, and a high (46 percent) mortality rate. After a gestation period of eight and one half months, a single infant weighing 4 lb (2 kg) is born that will develop at twice the rate of a human infant. Females reach maturity at eight years of age, while males remain sterile until they acquire their silverback, around eleven to thirteen years of age. It is the female that initiates mating, often very actively.

History

At one time, gorillas were much more widespread. During the last ice age, the disappearance of a large portion of the rain forest separated by more than 620 mi (1,000 km) the populations of East Africa from those of the West Africa.

Observation Site

Congo, Rwanda, and Uganda: The Virungas Chain and Bwindi Forest

These two high-altitude sites located in the heart of central Africa, are the sole home of the entire current population of mountain gorillas, perhaps the most endangered of all the great apes.

True "islands" of nature surrounded by farms and human installations, and numerous in the very fertile land of this region, these two mountain forests had their area considerable reduced throughout the twentieth century. Fortunately, rather effective conservation measures were finally introduced beginning in the 1970s, somewhat improving the situation and the future of the gorillas. The Virungas forest, which extends into the Congo, Rwanda, and Uganda, and the Bwindi Forest, which is completely situated in Uganda, are at least 62 miles (100 km) from each other. The importance of Bwindi in the conservation of mountain gorillas was recognized only recently; although since the 1930s the government authorities in the Congo and Rwanda (Belgium) created the first African national park in the Virungas (Albert National Park), it was necessary to wait almost 50 years more before Bwindi Forest—often called "the impenetrable forest"—was elevated to the rank first of nature reserve, then national park.

A dispute divided taxonomists for several years over the subject of the exact classification of the gorillas of Bwindi Forest; morphological details caused some experts to believe that they were a distinct subspecies of the animals living in the Virunga Mountains. Although these arguments among specialists have not been totally resolved, it seems certain today that these gorillas are part of the same population that may have divided only recently.

In 1996 the population of gorillas in Bwindi was estimated at between 300 and 320 animals, while almost 400 individuals were estimated to be in the Virungas Chain. Unfortunately, the tragic political situation that this magnificent region has experienced in recent years has considerably reduced the effectiveness of conservation measures. The violence has endangered tourism centered on the observation of gorillas, which had become one of the primary sources of income for these countries and one of the best guarantees for the conservation of these

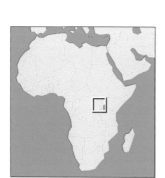
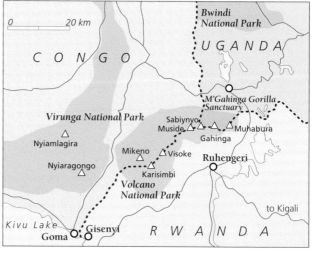

great primates. We therefore cannot recommend that anyone visit these sensitive areas while the political situation remains unresolved. Although many births have been recorded in the best-known gorilla groups since the beginning of the conflict, more than a dozen gorillas have also died violently during the same period.

Fauna and Flora

The flora of the Virungas and of Bwindi is markedly similar since the natural environments are practically identical. The Bwindi Forest, however, offers the advantage of having been less disturbed by farmers; the national park is the only natural site in this area that extends, without interruption, from approximately 3,608 feet (1,100 m) up to almost 8,528 feet (2,600 m) of altitude. Many types of vegetation, specifically great, dense expanses of bamboo, are better represented there than in the Virungas. The lower limit of the Volcano National Park, in Rwanda, is located at more than 7,872 feet (2,400 m) in altitude; on the other hand, the volcanoes there rise to more than 3,120 feet (4,000 m)—approximately 15,416 feet (4,700 m) for the highest, Karisimbi. This has made possible the development of combinations of vegetation typical of the African high mountains, such as the Afro-alpine prairies, or the open areas of lobelias and giant groundsels. These environments, however, are not good for gorillas; they rarely venture beyond the moist forest, which does not climb higher than 11,480 feet (3,500 m) maximum.

Bwindi is also richer in species of primates; while the only monkeys present in the Virungas—other than gorillas—are the diademed guenon, or golden monkeys, the impenetrable forest is home to nine other primates: a small population of chimpanzees, the red-tailed monkey, the diademed guenon, L'Hoest's guenon, the guereza, the olive baboon, and three nocturnal species—potto, Senegalese galago, and Demidoff's galago.

Among the other mammals present in the forest environments of the Virungas and Bwindi are the harnessed guib and the black-faced duiker—two forest antelopes—as well as the forest buffalo. A small population of forest elephants, severely endangered, survives in both sites, while the presence of the leopard have not been confirmed for several years.

Dian Fossey in the Mist

Dian Fossey, the famous primatologist who revealed to the world the tragic fate of the mountain gorillas, was born in California in 1932. After having read the works of George B. Schaller on gorillas in their natural habitat, she left for Africa in 1963 in order to see them up close. Once there, she met the anthropologist Louis Leakey. Believing that the study of the great apes could provide us with clarification about our own origins, Leakey had already encouraged Jane Goodall to study chimpanzees. He did the same with Dian Fossey for gorillas, and some years later with Biruté Galdikas for orangutans.

Dian Fossey began her study in Zaire in 1967, but soon moved to Rwanda, in Karisoke, where she established a research center. Little by little, she gained the confidence of the gorillas in the area, and came to know them individually. In 1983 she published her famous book *Gorillas in the Mist*, which told of her experiences among these peaceful, sociable animals. In 1985, after 22 years of life devoted to the study and survival of gorillas, Fossey was found murdered in her cabin. The murderer, probably a poacher, was never found.

Dian Fossey had contributed to destroying the image of a King Kong man-eater and to make known the incomparable value of this animal. But the battle has not been won. The mountain gorilla is still headed for extinction and remains perhaps one of the rarest mammals on earth.

It was in this cabin that Dian Fossey lived while conducting her research on gorillas.

Barbary Macaque

(Macaca sylvanus)

GERMAN: Magot
FRENCH: macaque de Barbarie
SPANISH: mono de Gibraltar
ITALIAN: magot

Description

Average length: male head and body between 21 and 26 in (53.5 and 66.5 cm): female head and body between 19 and 24 in (48 and 60 cm).
Average weight: adult between 11 and 22 lb (5 and 10 kg); newborn 16 oz (450 g).

Of average size, the Barbary macaque is a robust monkey. Its fur is thick, except on the stomach. The face is lighter than the body, with thick hair covering the forehead up to the eyebrows. The tail is reduced to a stump with no vertebrae. The side-whiskers are thick, and the chin adorned with a goatee. Infants are actually very dark, but lighten around six months.

Family

Cercopithecidae.

Locomotion

The Barbary macaque is one of the most terrestrial of monkeys. It moves on four feet.

Diet

The Barbary macaque feeds on leaves, sapwood, and seeds from the cedar tree, berries, greenery, buds, grass shoots, roots, and invertebrates. Its raids in Gibraltar are currently under control, but it can still, in North Africa, be audacious, coming down from the mountains to plunder fields and gardens.

Predators

It is hunted by man, but since lions and leopards have disappeared from North Africa, it no longer has any natural predators.

Longevity

20 years or more.

Distribution

The distribution of the Barbary macaque is currently limited to the Rock of Gibraltar and to the oak and cedar groves in the Atlas Mountains in Morocco and in the north of Algeria. This makes it the only primate present in Europe and the sole macaque in Africa.

Status

It is estimated that there are approximately 8,000 to 10,000 individuals in Morocco, and 5,000 in Algeria. The Barbary macaque is currently threatened by the destruction of its habitat. In fact, pastures for cattle, sheep, and goats prevent the reforesting of the land, originally denuded by logging.

Social Organization

Barbary macaques live in multimale troops. The males take particularly good care of the young. This common passion for babies can even be the reason for two rival males to meet amicably for a session of grooming a black infant, often in spite of its vigorous protests! It appears that babies also are used to reduce tension levels. During antagonistic encounters, males will present babies to aggressors, who then begin grooming the infants. The large number of males in Barbary macaque groups may have fostered this behavior. In the event of dan-

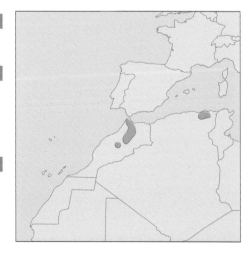

ger, it is the entire troop that protects the young. At twilight, macaques vocalize together, apparently both to indicate their number to potential predators and to establish the proximity between monkeys in the dormitory. Females, infants, and juveniles sleep in groups of two or three individuals, the composition of which varies from night to night, making it possible for all to better know each other and to reinforce the bonds of the entire troop. Males generally sleep alone or in the company of juvenile males.

Behavior

The mother of a newborn female spends her time in the company of other members of her own maternal line, while the mother of a newborn male frequents the members of other lines. At the end of five months, this process of socialization yields conclusive results; the young females remain strongly linked to the family network, while the young males integrate themselves more widely into the troop. This strategy is adopted because females form the permanent core of the social unit. The young females stay within the natal group whereas the males generally emigrate at puberty or shortly thereafter.

Reproduction

Females appear less particular than males in their choice of sexual partners, simply preferring their "friends," regardless of rank. On the other hand, older males prefer to mate with females of high hierarchical rank, leaving the females of lesser ranks to the younger males. After a gestation period of approximately seven months, one or two infants are born, most often in July and August.

History

At one time, the Barbary macaque, probably the close descendant of the macaque that went on to populate Asia, was spread throughout the Mediterranean. It was the monkey that was most familiar to the ancient civilizations of the West. It survived in Spain in a natural manner until the 1890s.

Male macaques are very involved in the life of the youngsters.

Observation Site

Gibraltar: Monkey Hill

A British possession for centuries, the famous Rock of Gibraltar rises abruptly at the southernmost tip of Spain, opposite the African coast, some fifteen kilometers east of the city of Algeciras.

The history of the Barbary macaques of Gibraltar has become almost legendary. Probably introduced around the mid-eighteenth century, these animals are the only primates leading a totally free life on the European continent. There is a widespread superstition that if the macaques disappear from Gibraltar, the United Kingdom will quickly lose its supremacy over the celebrated rock and the strategic position that it affords.

Apparently conscious of this unique status, the macaques of Gibraltar have long behaved like despots, not hesitating to venture into urban areas below the hills to engage in plunder (pillaging food in markets, gardens, and so on). At present, measures have fortunately been taken in order to prevent such behavior, which is as insolent as it is undesirable.

Fauna and Flora

Most of the Rock of Gibraltar is covered with scrub vegetation in which the wild olive tree (*Olea europea*) dominates, and where small shrubs and thorny plants abound. Pines (primarily the Alep pine) are sprinkled over the hills, occasionally forming scattered forest areas.

Macaques are the only mammals easily visible on the Rock; birds, on the other hand, are a great attraction during migratory periods. Gibraltar has become a major observa-

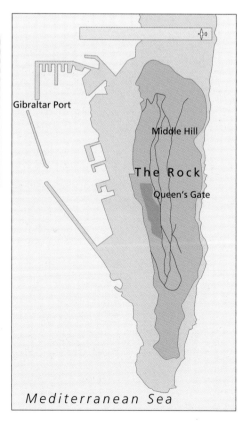

tion site during migration, particularly of raptors, attracting thousands of ornithologists each fall.

It is also a well-known site for observing dolphins, which regularly cross the strait in large schools. Many tourist operators specialize in tours to view these marine mammals, and organize daily excursions that occasionally make it possible to admire several hundred common dolphins, bottlenosed dolphins, and the like.

Observation

Observing the Barbary macaques of Gibraltar is particularly easy, as the animals have, in fact, lost all fear of man, and often move about among visitors on the slopes of the Rock. Two distinct troops inhabit the site permanently—that of Queen's Gate and that of Midden Hill. Since 1972 only the first is accessible to visitors, the other colony being located in military territory, where access is not permitted.

Gibraltar had no more than 40 macaques in the 1970s, but the population today is slightly over 100 individuals. A plan for a national park specifically dedicated to the Barbary macaques and called Monkey Hill has been proposed to the local authorities for the last 25 years; right now, although officially approved, it exists only on paper.

It must be remembered that macaques—despite appearances—are first and foremost wild animals, and visitors should behave accordingly. Avoid giving them anything to eat, and try to keep a big enough distance to not encourage close contact (such as pilfering of bags, and so on), and to avoid accidents such as biting, which unfortunately take place each year.

A confirmed acrobat, the Barbary macaque (seen here, a female) is perfectly adapted to human structures on the famous Rock of Gibraltar.

The Conquerors of Gibraltar

Today, the macaque population of Gibraltar is maintained thanks to the contribution of monkeys from the Atlas mountains; however, their origins in the strait remain obscure. Perhaps they are the vestige of populations that formerly covered Mediterranean Europe, but according to some historians, it was the Carthaginians or the Romans that may have left individuals there that originated in North Africa. In any event, they were already there in 711, when the Arab conqueror Tarik Ibn Sijad, for the first time, crossed the strait that would bear his name (Gibraltar means "Tarik's rock").

It was not until 1856 that the macaques appeared in the official documents of the English army, when the British governor issued a decree protecting them. The macaques lost no time in dominating the city, pillaging houses, stores, and gardens, strangling chickens, and striking women and children. However, the current measures that relegate them to a restricted area were taken only when a macaque helped himself to the plumed helmet of the governor in order to "ape" His Excellency on the battlements of the fortress.

Winston Churchill illustrated the force of the tradition, according to which, the day the monkeys disappear, the English will lose the fortress. During the summer of 1942 he sent a telegram to the chief of the British military forces in North Africa ordering him to immediately capture monkeys for Gibraltar!

The fate, like the history, of the Barbary macaques is closely linked to that of man.

PRACTICAL INFORMATION

Tourist travel to the site during the summer is often *very* dense.

TRANSPORTATION

Since the total opening of the border with Spain, it is easy to visit the city, the port, and the Rock in a single day. It is, however, preferable to take at least two days, in order to visit the Rock in a relaxed manner.

■ **BY PLANE.** Regular departures from Great Britain on Gibraltar Airways leave almost every day.

■ **BY CAR.** Access to the enclave is very easy by road—National Road 351 leaving the town of San Roque the length of the long coastal road connecting Algeciras with Malage via Marbella.

DISTANCES

Malaga to Gibraltar: 56 mi (90 km).

Gibraltar to Algeciras: 9.3 mi (15 km).

ACCOMMODATIONS

An impressive number of establishments make it possible to stay in Gibraltar: inns, hotels, rooms with residents, and others. There are possibilities for all budgets. It should be noted that the city can be rather noisy for those not accustomed to Mediterranean night life.

CLIMATE

The climate in Gibraltar is a typical Mediterranean climate, and often corresponds to the conditions prevalent in Andalusia—winters are generally mild, summers are dry and hot. In the city, the heat can become torrid at midday. In the heights, the air becomes more breathable, but caution must be taken with the sun, which is particularly intense in the open rocky areas.

TRAVEL CONDITIONS

Access to the Rock and to Monkey Hill is free and it is authorized between sunrise and sunset all year. Numerous paths make it possible to climb the hills, particularly in the area most frequented by the macaques of the Queen's Gate group. A cable car (for a fee) makes it possible to reach the top of the Rock.

Chimpanzee

(Pan troglodytes)

GERMAN: Schimpanse
FRENCH: chimpanzé
SPANISH: chimpancé
SWAHILI: soko (mtu)

Description

Average length: male body plus head 33 in (85 cm); female body plus head 31 in (77.5 cm).
Average weight: male 88 lb (40 kg); female 66 lb (30 kg); newborn 4.2 lb (1.9 kg).

This is a large monkey, whose dark coat makes it possible to see the pale skin of the body. The face, with its large, protruding eyes, is generally pigmented, occasionally with white hairs on the muzzle. Youngsters have a tuft of white hair on the haunches. The two sexes have a tendency to become bald rather early, especially the females. Unlike other great apes, the male's genital organs are very visible.

Family

Pongidae.

Locomotion

The chimpanzee sleeps in the trees at night and feeds in them during the day, but generally, it travels on the ground. Less acrobatic than its close relative, the bonobo, the chimpanzee does engage in brachiation, and climbs well. On the ground, it moves by leaning forward on the external face of the phalanges, but occasionally walks upright.

Diet

Primarily a fruit eater, the chimpanzee also eats nuts, grains, and young leaves. It also hunts young colobus or baboons, often alone, occasionally in groups.

Predators

Leopards, rock pythons, man.

Longevity

50 years in nature.

Distribution

In the west, the center, and the east of Africa, as far as

A male chimpanzee has the strength of three men.

Chimpanzee

Tanzania and Zambia in the south. Previously, it was thought that chimpanzees lived only in the forest; now it is known that they can also tolerate the dry conditions of the wooded savannah.

Status

The chimpanzee is endangered. Estimates vary but, at the present, there are approximately 100,000.

Social Organization

The chimpanzee lives in multimale troops of 30 to 80 individuals that disperse in small bands of varied and changing composition. The hierarchy between males is determined by displays during which they march on two feet, shake branches, and throw stones while howling, hair raised, giving them a formidable look. This hierarchy is complicated by deep friendships. Males are, however, always dominant in comparison with females. The youngsters, with their still white haunches, are lords and masters of everything.

Vocal and visual communication is as varied as it is important among chimpanzees. Facial mimicry and postures are particularly rich. Most are familiar to us, such as the smile or sulking, or the gestures that accompany meetings after long separations, and are expressed according to the respective status of each, ceremonially for some, affectionately for others that throw themselves into each other's arms, hugging and patting each other on the back.

Other forms of physical communication, like the showing of the haunches to hierarchical superiors, are strange to us.

Behavior

In spite of its short thumbs that prevent a good grasp, the chimpanzee constructs nests and knows how to alter natural objects to turn them into tools. Sometimes it even uses a tool "kit"; for example, it will extract sap with the help of a mortar, then sponge it with natural fibers.

Reproduction

Very similar to that of man. Heat occurs every 35 days, and is signaled by the edema of the skin on the genitals, which become pink. Gestation lasts 228 days. Giving birth is easier than in humans, as the newborn is smaller compared to the size of the pelvic basin. Twins are rare. Weaning is completed at five years. Females reach puberty at eight to nine years, and give birth to their first infant at eleven to twelve years. Males do not reach their adult size and social maturity until fifteen years of age.

History

Some molecular data lead us to believe that not only are chimpanzees our closest relatives, a widely recognized fact, but that we are also theirs. According to one study, not yet concluded, we have shared the same branch of our genealogical tree for three million years *after* the divergence of the other great apes. That means that chimpanzees are no closer to a gorilla or an orangutan than we ourselves are.

Observation Site

Tanzania: Gombe Stream National Park

Gombe Stream National Park is found in the extreme west of Tanzania, between the border with Burundi in the north and the port city of Kigoma in the south, approximately 9.3 mi (15 km) from the latter. It runs along the eastern shore of Lake Tanganyika for several kilometers and goes as far as the top of the high hills that tower above it.

Gombe Stream is known throughout the world thanks to the work of the American researcher Jane Goodall. Her studies on chimpanzee behavior in the natural environment are the longest that have ever been conducted on this species. Arriving at Gombe Stream in 1960 on the initiative of the celebrated Dr. Louis Leakey, Jane Goodall has, in fact, continued her observations in an uninterrupted manner for more than 35 years! The results of her research have provided an impressive quantity of information on the ways of chimpanzees, their social relationships, their prodigious capacities for adaptation, and so on; it is, in fact, Jane Goodall who discovered various important characteristics of their behavior, such as the conscious and well-thought-out use of objects, the associations and the intrigues between individuals of the same line, the actual tactics of war used by one "tribe" against a competitor, cannibalism, and other facts. It should be noted that the chimpanzees in the park belong to a subspecies typical of the wooded savannahs of Eastern Africa, and are occasionally markedly different from the individuals that can be seen in Western Africa.

Gombe Stream National Park is a small park, by Tanzanian standards, at least. Its relatively reduced size is a handicap for the

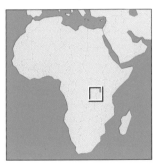
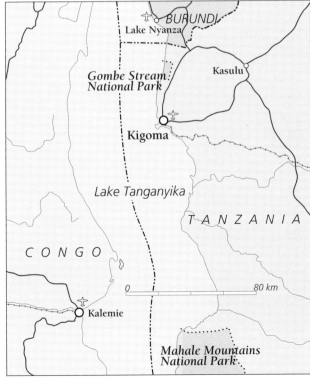

long-term survival of rather large communities of animals within an area with a human population that is becoming more and more dense.

Fauna and Flora

The park is characterized by wooded areas that are often relatively homogenous, where several species of acacias dominate, but there are also hills that are open and covered with high herbaceous vegetation, regularly ravaged by brush fires from the outside. Along the length of the streams that flow down the hills, or that have their source there, as in the more enclosed valleys, there are gallery forests that are clearly more tropical, and that

Scientist Jane Goodall has for several decades been engaged in an in-depth study of chimpanzee behavior.

These lush mountain landscapes washed by the shores of Lake Tanganyika in Tanzania are home to a national park created especially for the safeguarding of chimpanzees: Mahale Mountains Park. Difficult to get to, it receives few tourists.

contain, among others, imposing fig trees that are often quite spectacular.

Primates are probably the principal attraction of Gombe Stream. In addition to chimpanzees, the park contains a substantial population of baboons. Other primates that can often be seen are the crowned or "blue monkey," the vervet monkey, and the coloba bai, as well as the galagos.

The forest galleries are the domain of antelopes such as the harnessed guib and various duikers, while the warthogs prefer more open spaces. Birds are also numerous in the park.

The prepared ornithologist will recognize the usual species of the wooded savannahs of Eastern Africa and others more typical of central Africa and the great tropical forest, not to mention the species that favor aquatic environments that find favorable locations along the lakeshore.

Regardless of specific interests, one should plan to check various habitats at different times of the day and night, so as to maximize the number of species encountered.

Observation

Gombe Stream National Park is incontestably among the best sites in the world for the observation of chimpanzees in their natural environment. Many groups have, in fact, become habituated to the presence of humans for decades, and do not show any fear at the arrival of those strange primates called humans. When the chimpanzees are near, it is important to follow instructions regarding proper behavior, as failure to do so can be dangerous to both the chimpanzees and the visitors.

A feeding station, where food is left for monkeys, has been in operation since the first stages of Jane Goodall's research; the observation conditions here are thus not very "natural." By careful observation, however, one can see interactions and behaviors not observable

Apprenticeship or APErenticeship?

The idea of language in animals has always been the subject of lively debate in the fields of psychology, biology, anthropology, and philosophy. According to Descartes, the word is the sole evidence of the presence of thought. Many contemporary authors still share this opinion, but must this word be spoken by the mouth?

According to the researcher Sue Savage-Rumbauch, it is extremely anthropocentric to imagine that the great apes would use language in the same way that we do, as their physiology prevents them from producing the same sounds. On the other hand, as Darwin has suggested, it can be reasonably thought that if we speak, we must have inherited the bases for language from our ancestors, bases that the great apes, too, would also have inherited.

Research seems to head in this direction. In fact, the same portion of the brain controls movements of the hand and the learning of language. This explains, in part, the astonishing results obtained for the last 30 years in teaching sign language to the great apes. The scientist Roger Fouts (author of *The School of Chimpanzees*) and the guenon Washoe, who possesses a vocabulary of approximately 1,000 words, were the first to reveal the ability of the great apes to associate abstract words in a complex order, which itself conveys meaning. They distinguish between proper and common nouns, use words in the absence and in anticipation of their referents, classify them in categories, create new words, and so on.

Of course, this language has its limits, but other than the simple manipulation of words, the great apes also use signs to speak of tears, of friends, of lies, and of love.

Chimpanzees in the wild use various gestures to signal, for example, a stop or an invitation for grooming. These gestures vary with the population, with the juveniles learning the "dialect" of their own.

under other circumstances, and a great deal can be learned. Obviously, the contact is "staged," but such is the trade-off necessary if one wishes to observe with certain degree of reliability these normally wary animals.

A guide is obligatory in order to search for chimpanzees on foot. In principal, each group of visitors has the right to remain with the animals for a maximum of one hour.

For real chimpanzee fanatics, a visit to another national park created primarily for the safeguarding of these primates in Tanzania can be very interesting. This park is in the Mahale Mountains, also located along Lake Tanganyika, but approximately 124 mi (200 km) south of Gombe Stream. Clearly larger—approximately 395,200 acres (160,000 ha), this park of high hills and mountains that jut into the lake is home to approximately 700 chimpanzees. A group of these animals has been studied by a team of Japanese scientists since 1961, almost as long as at Gombe Stream. For obscure reasons, the lack of media interest has affected this team, as their work is known only in the scientific world, unlike that of Jane Goodall. Access to this park, however, is not easy—chartering a small carrier plane in Dar-es-Salaam (later Kigoma) is the only possible option for a short stay at Mahale Mountains, where no road provides access.

PRACTICAL INFORMATION

Our summer is the best time of the year to visit Gombe Stream, the dry season.

TRANSPORTATION

Tanzania is linked to the major U.S. airports by regularly scheduled flights (more frequent to Dar-es-Salaam, and less frequent to Arusha).

■ **BY CAR AND TRAIN.** Kigoma, the preferred access route to Gombe Stream, can be reached either by road (via Mwanza), or by train (departing Dar-es-Salaam, via Dodoma).

■ **BY BOAT.** Gombe Stream can only be reached by boat, as no road or trail runs the length of the eastern shore of the lake, with its very hilly relief. Boat taxis operate somewhat regularly between Kigoma and the border town of Nyanza-Lake, in Burundi; they generally depart at the end of the morning from Kigoma. This method is inexpensive but requires adaptation to African rhythm. Boats stop, in fact very regularly, to take on or let off passengers along the shore, operations that sometimes seem to take an eternity to Westerners! However, the current precarious political situation in Burundi makes the option of reaching Gombe Stream by departing from this country risky. The best option for visitors wishing to reach Gombe Stream remains chartering a tourist boat from Kigoma. Various agencies provide this type of service, which is rather expensive.

ACCOMMODATIONS

There is a tourist campground in the national park in a very pleasant area on the shore of Lake Tanganyika. Bungalows (*bandas*) are available for visitors; they each contain several rooms of rather basic comfort. Rudimentary sanitary facilities are also available, as is a common kitchen. All food must be brought in, as nothing is available in the park.

It should be noted that baboons, which are numerous in the park, are particularly bold and can represent a very real danger. Keeping food in your tent is, in fact, suicidal in Gombe Stream!

CLIMATE

The climate is generally very pleasant: 77–86°F (25–30°C) during the hottest hours in the rainy season (March to June and October to November), and somewhat lower in the dry season (July to September and December to February). Since the park is located at a high altitude, the nights are sometimes rather cool so you are advised to bring something warm for the evening.

The area is drenched during the major rainy season (March to June), but storms can occur throughout the year, generally during the afternoon or in the evening.

TRAVEL CONDITIONS

Admission is charged for entry into Gombe Stream National Park and is rather expensive. There is no chance of staying outside of the protected area.

All travel is on foot, and the park has neither road nor trail. The hilly terrain and the heat can make walking difficult for those who are not accustomed to physical exertion. A guide is obligatory for traveling in the park.

Gelada
(Theropithecus gelada)

GERMAN: Dschelada
FRENCH: gélada
SPANISH: gelada
ITALIAN: gelada

Description

Average length: male body plus head 28 in (70 cm), tail 20.3 in (51.5 cm); female body plus head 22 in (55.5 cm), tail 18 in (46.5 cm).
Average weight: male 46 lb (21 kg); female 31 lb (14 kg).

Unlike baboons, with which they were formerly classified, the gelada has nostrils that open sidewards; it is actually a separate genus. Instead of ischial callosities, the gelada has fleshy cushions, making it possible, like the thighs in man, to sit comfortably and for as long as it wants. Its coat is brown or blackish, with the end of the tail tufted. Males have a coat of long, thick hair as well as prominent whiskers. On the throat, both sexes have a patch of naked skin in the shape of an hourglass. This is edged with white hair in the male, and peaked with prominent carbuncles in the females, which cover the perineal region. These conspicuous areas of skin are visible over long distances in the gelada's open habitat.

Family

Cercopithecidae.

Longevity

In captivity, 20 years.

Distribution

The gelada is found only on the prairies of the high plateaus located between the deep gorges in the north and center of Ethiopia, at altitudes varying from between 7,708 and 14,432 ft (2,350 and 4,400 m). They are never found more than 1.2 mi (2 km) away from the gorges where they take refuge in the event of predators.

Status

Fortunately, the gelada is not yet an endangered species, but its habitat is giving way to agriculture. In the southern part of this territory, the male is hunted for his magnificent coat, which is used in making headdresses and capes.

The gelada is also threatened by hybridization with the olive baboon.

Locomotion

Living in the plains of high plateaus almost entirely without trees, the gelada is the most terrestrial of primates, outside of man.

Diet

This is the only primate that is entirely graminivorous, and its large molars are well adapted to this diet. The gelada collects the grass blade by blade, waiting until it has a fistful before bringing it to its mouth. It detaches the seeds by sliding the stalk between its teeth or between the thumb and index finger. By using two hands, it quickly unearths rhizomes. The gelada moves very little while eating, dragging its padded thighs from one place to another without getting up.

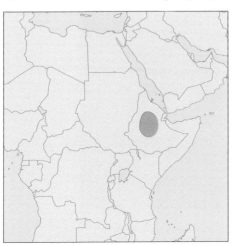

Predators

Man, leopards, dogs.

Social Organization

The herbivorous diet makes possible the formation of large troops of up to 400 geladas that sleep together on the cliffs. If food is abundant, the troop can hunt it together during the day; if not, geladas disperse in small bands of up to 20 members, with a single male, which appears to organize movement. In this

Gelada

way, he attempts to bring his females together, although the actual social cohesion of bands is due more to the very strong bonds between females. They participate in defending the troop and can, on occasion, unite against the male.

Behavior

Approximately one hour after sunrise, troops emerge from the gorges onto the plateaus. Two hours are spent in social activities, before departing in search of food, around 9:00 to 10:00 A.M. At 4:00 P.M. geladas return to the cliffs, where they spend the night.

Geladas have a repertoire of approximately 25 vocalizations, with sounds comparable to our consonants and vowels. They have a unique behavior for expressing annoyance, greeting, or friendship, consisting of pulling back the upper lip, which rolls back above the nose.

Reproduction

The menstrual cycle lasts 32 to 36 days. Heat is signaled by the edema and reddening of the vesicles surrounding the naked skin of the chest. This manifestation is perhaps due to the fact that the perineum is hidden by the quasi-permanent sitting position that characterizes the species. The teats, scarlet during lactation, are side by side, making it possible for the infant to take both in its mouth at the same time.

History

The tooth of a gelada dating from four million years ago has been found. Most of the fossilized bones of geladas have been found at the same time as the bones of the hippopotamus, on the shore of the lake. In the Pleistocene era, geladas ranged from Algeria to the Cape of Good Hope, and one of its species reached the size of a female gorilla.

Geladas spend the night in the security of the cliffs. In the foreground, a male, recognizable by his long side whiskers.

Observation Site

Ethiopia: Simian Mountains National Park

The Simian Massif, located in northern Ethiopia, approximately 62 mi (100 km) north of the town of Gonder, is one of the highest regions in this large country; most of its mountaintops are higher than 13,120 ft (4,000 m). The highest point in the country, Ras Dashen—15,154 ft (4,620 m) is found here.

Simian Mountains National Park extends across the entire western portion of this great mountainous range, over approximately 44,460 acres (18,000 ha). It consists of several large plateaus of high altitude, crosscut by numerous, often abrupt valleys. In the north and east of the park, an escarpment of almost 3,280 ft (1,000 m) in height plunges without stop toward the immense plains that continue north as far as Eritrea.

The park ranges between 6,232 and 13,120 ft (1,900 and 4,000 m) in altitude, with various stages of distinct vegetation dominating below the land cultivated for agriculture. In the lower levels of the park, there are still tropical forests, seriously endangered in Ethiopia, while in the higher levels, the forest quickly gives way to shrub vegetation where species such as the *Hypericum* and the lobelia dominate, reaching surprising size. The highest levels are characterized by an Afro-alpine mountain environment with low vegetation.

Practically isolated from the outside world for several decades following the terrible civil war that ravaged it, Ethiopia has known peace for several years, and tourism—very limited, at present—is developing once again in this immense country, whose natural heritage, along with its culture and history, are among the most remarkable in Africa.

Fauna and Flora

Almost 100 species of birds frequent the park or live there permanently, not a high number in comparison with other areas in the horn of Africa, but that is explained by the limited botanical diversity in the mountainous regions very specific to this part of Ethiopia. The high cliffs that border the most enclosed valleys of Simian are, however, favorable for

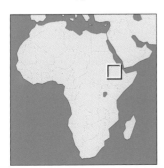

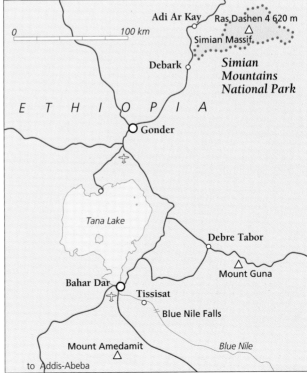

birds that establish their colonies in this type of environment, among them the bearded vulture, a superb type that is particularly rare.

Large mammals include various species that are rare or that have almost disappeared everywhere else. This is the case with the mountain nyala, an antelope whose numbers are few in all of the regions where it is present. Among other ungulates worthy of note are the Klipspringer, a small, elegant antelope that is particularly agile, inhabiting the rocks and cliffs where it moves about with great ease, and especially the Walia ibex, one of the rarest mammals on the continent. Thanks to adequate protective measures, its numbers have more than tripled in the park in the last 25 years, reaching approximately 500 today. The harnessed guib and Grimm's duiker are among the other animals it is possible to see in the park.

As for the canidae, in addition to jackals, it is the Ethiopian wolf that attracts attention. This African wolf, classified as a grey wolf of the same species as other Eurasian populations, exists only in the Ethiopian highlands, and specifically in Simian Mountains National Park, where its population is estimated at about 100 individuals. The simian jackal is one of the world's rarest canids (wild dogs). Less than 50 remain at Simian Mountains, with perhaps 500 more to be found in the Bale Mountains. Its fur is a beautiful tawny rufus color, with white chin, ears, and underparts. The population has declined because of agricultural development, and because the jackal is hunted by farmers who falsely believe that it kills sheep.

The gelada baboon is among the most spectacular mammals in the national park. Living in large groups, this monkey, with its remarkable coloring, numbers approximately 20,000.

Observation

The gregarious manners of the gelada—its troops can contain several hundred individuals—as well as the characteristics of its environment (open areas of low vegetation), make this species easily observable in Simian Mountains National Park; however, no troop is actually habituated to the presence of visitors in the park, which is rarely visited at present. And while

Geladas in their natural environment, in Ethiopia. The gelada is the most mountainous of monkeys and certainly the most terrestrial; the highest plains are its environment of choice.

it is sometimes possible to surprise them or to get relatively close to them, their escape route remains very large.

Guides generally know the areas where they can be found, and will escort those wishing to go. The park's administrative center is located in the western part of Sankaber.

PRACTICAL INFORMATION

The best time of year to visit the National Park is between December and March.

TRANSPORTATION

Addis Ababa, capital of Ethiopia, is linked to Los Angeles, Washington, and New York by Ethiopian Airlines. The quickest option—and the most realistic for a short stay—for reaching Gonder from Addis Ababa is to rent a vehicle.

■ **BY TAXI.** Departing from Gonder, it is possible to reach Debark, the town closest to the main entrance of the park, either by using local buses or by hiring a taxi, a quicker solution since it takes only about two hours to travel the distance between the city and the park's main entrance.

■ **BY CAR.** The most comfortable option is, of course, to rent a car, which is, however, particularly expensive in Ethiopia. There are only a few tour operators scheduling trips to Simian at present, but this situation will probably improve in the future.

ACCOMMODATIONS

A single small hotel, of minimum comfort, currently exists in Debark; otherwise, one must go to Gonder, where there are other lodging possibilities.

In the park, the only solution is camping, in almost total autonomy. Generally, it is possible to find a hut belonging to one of the inhabitants in the villages surrounding the park. In any event, it is a good idea to remember the tragic situation from which Ethiopia is slowly emerging, and not to expect to find tourist infrastructures or hotels comparable with those in other tourist countries, such as Kenya, for example.

CLIMATE

It is generally sunny, but rain can also be frequent and often rather violent. From April to September, the rains can make travel difficult, and fog often covers the countryside; October, November, and December are the coldest months. The average daytime temperature ranges from 52–65°F (11–18°C) throughout the year; temperature decreases in relation to the altitude, and nights are always cooler in the park. Above 9,840 ft (3,000 m) night frost is very common, particularly during the dry season.

TRAVEL CONDITIONS

At present, the number of tourists in Simian Mountains National Park remains very limited.

A visit permit (for a fee) must be obtained at the Animal Service Bureau located in Debark before being allowed to enter the protected zone. A guide must accompany visitors at all times. It is better to arrange for the permit and the guide several days before the visit, possibly from the Animal Service Bureau in Addis Ababa.

Hikes lasting several days can be organized in the National Park, and are unquestionably worth the trouble for those able to do without comfort, the effort being largely compensated for by the vision of the sumptuous Simian landscape. Overnights are spent camping in the wild. For hikes that diverge substantially from the areas near the main entrance, one or several armed guards (rangers) must escort visitors, for reasons of security, in order to prevent an attack by bandits who may still come down from the mountains; although this has actually occurred in the recent past, their number has decreased.

All of the fees connected with a stay in the park for the guide and the rangers are the visitors' responsibility. Perfect autonomy is also required with regard to food and equipment, as no accommodations are currently available in the park.

It is difficult to find porters to carry the equipment and food while hiking, but it is easy to rent mules.

OBSERVING LEMURS IN MADAGASCAR

Madagascar, the Island Continent

About 165 million years ago, the Island of Madagascar was still part of the African continent, from which it is now separated by the Mozambique Canal, 249 mi (400 km) wide. This gigantic island—its surface area is twice the size of Arizona—possesses a biological and geological diversity that has earned it the name of the "island continent."

If we consider that this island was part of Africa—which is not the opinion of all specialists—it represents less than 2 percent of its area. Nevertheless it is home to one-quarter of all plant species—1,000 species of orchids, and the famous baobab, symbol of Africa, which exists in only one species elsewhere, exists in six here. About 73 percent of its mammals, including all the lemurs, are endemic. Madagascar chameleons represent two-thirds of the species throughout the world. The same is true for birds and insects, particularly butterflies, which attract admirers from all over the world. Madagascar is thus considered a conservation priority by the international community. Monkeys never came to Madagascar; the island had separated from the continent prior to the appearance of the first mammals. Some arrived there, perhaps on board rafts made from plant debris

carried by the currents. Among them were the first lemurs, which strongly resembled the mouse lemurs, the smallest of the present-day lemurs, but also the smallest primates. These migrations ended approximately 40 million years ago, before the appearance of monkeys. Lemurs, without the competition from simians, were thus able to diversify and prosper. The arrival of man on the island—the first, from Indonesia, then from India and Africa, landed 2,000 years ago—resulted in the extinction of eight genera of lemurs, consisting of some fifteen species, with the largest the size of a male gorilla—352–440 lb (160–200 kg). The remaining species are still quite varied and today number around 30.

The very colorful *Cameleo parsoni* is found in the eastern forests of Madagascar.

Lemurs can be divided into five families:

The **Lemuridae** (lemurs or makis) differ in their coat, their behavior, and their habitat, but they share the same profile, with a muzzle similar to that of a fox. These are the lemurs that are currently seen in captivity.

The **Cheirogaleidae** include the various species of mouse lemurs and dwarf lemurs, small nocturnal lemurs whose eyes glow like headlights wherever you walk in Madagascar at night.

The **Indriidae** include the indri, sifaka, and avahi. These animals hold themselves vertically in the trees, where they are capable of powerful jumps. They feed almost exclusively on leaves.

The species belonging to the **Megaladapidae** or koala lemurs and weasel lemurs appear in a wide variety of

The Perinet Reserve is part of the eastern mountain range that extends from north to south across the major portion of the Island of Madagascar.

habitats, from pine forests to damp forests. These nocturnal animals are easier to find by the strong vocalizations of males than by searching for their small round heads.

The family **Daubentoniidae** includes only one species, the aye-aye. Its bizarre appearance formerly caused it to be considered the ghost of departed relatives. While this belief, which protected it from man, is disappearing, the fear that it inspires continues, and today it is often killed on the spot.

PRACTICAL INFORMATION

The months of January through March are the worst time to visit Madagascar, where the rains can be a real obstacle to travel. The rains are less frequent between May and September.

Two sites make it possible to have a good idea of the various Malagasy habitats and species: the Perinet Reserve in the eastern part of the island and that of Berenty, in the south.

TRANSPORTATION

Two weekly flights are provided by Air France from the United States, with embarkation points in New York, Boston, Los Angeles, Chicago, and Houston. Daily flights (except for Wednesdays) link La Réunion with Antananarivo.

CLIMATE

Madagascar has a climate that is tropical and moist; the heat, however, is more moderate in the mountainous areas than on the plain; nights can be relatively cool. The rains are more intense there.

Perinet Reserve

The special reserve of Perinet-Analamaotra (also called Andasibe) is located midway between Antananarivo and Tamatave. It encompasses the eastern mountain ridge that extends north-south across the major portion of the island of Madagascar and consists primarily of moist forest. It is celebrated for being home to a substantial population of indris, the largest living lemur. Several dozen family groups of this remarkable and engaging animal actually live in the reserve, totaling more than 250 individuals. The best time to see them under good conditions is in the early morning and the end of the afternoon, when the heat is less intense and the indris more active. It is also at the beginning of the day that it is possible to hear the mysterious cry of these lemurs, which has been compared by some to the legendary call of the sirens.

In addition to the indri, at least five other species of lemur populate the reserve; those that are most likely to be observed are the lesser gray bamboo lemur, the woolly lemur, and the red-fronted brown lemur. The reserve is also home to the famous and surprising aye-aye, as well as a dozen species of tenrecs, an avifauna represented by more than 100 species of reptiles and frogs, including the spectacular Parson's chameleon, whose length can exceed 12 in (30 cm).

ACCOMMODATIONS

Various possibilities for lodging in relatively basic comfort are available in Andasibe. It is also possible to camp near the official entrance to the Perinet Reserve. For those who would like a bit more comfort, the capital remains the best solution, with the distance to the reserve not very great.

Berenty Reserve

Approximately 62 mi (100 km) to the southwest of Fort-Dauphin (Taolanaro), in the extreme south of the island, is the private Berenty Reserve. Here, the thorny forest is populated by amazing plant species, including the spiny plants that belong to a family endemic to Madagascar, the Didierea, as well as numerous euphorbias of varied shape. Well protected, this small reserve is astonishing for the concentration of fauna that it houses. Well-maintained paths make it possible for the visitor to observe lemurs, mouse lemurs, and dwarf lemurs without difficulty. The reserve is also home to a large number of frogs.

ACCOMMODATIONS

It is advised that visitors remain on site in order to observe the animals at night and in the early morning. Some 30 comfortable bungalows have been made available for this purpose. A small ethnological museum makes it possible for visitors to stay out of the sun during the hottest hours of the day.

Indri
(Indri indri)

GERMAN: Indri
FRENCH: indri
SPANISH: indri
ITALIAN: indri
MALAGASY: endrina or babkoata (pronounced "babacotte")

Description

Average length: adult head and body 28 in (70 cm).
Average weight: adult 15–22 lb (7–10 kg), sometimes more.

The indri is not only the largest of the lemurs but also the largest of the prosimians. While its black and white coat, thick and silky, resembles that of the lemur *Varecia variegata*, it is distinguished by its tail, reduced to a simple stump, and its vertical position in the trees.

Family

Indriidae.

Locomotion

The indri is completely arboreal, preferring large diameter vertical trunks on which it moves by leaping.

Diet

It feeds principally on young leaves, shoots, flowers, and fruits. Females and infants always have priority during frequent conflicts around food.

Predators

Raptors are less of a danger than deforestation or hunting, even if the taboos surrounding the hunting of the indri are still fortunately in effect among certain ethnic groups.

As it floats over the top of the trees, the cry of the indri resembles the song of a whale.

Longevity

Unknown.

Distribution

The indri is limited to the rain forest of the eastern slopes of areas up to 5,904 ft (1,800 m) in altitude in northern and central Madagascar.

Status

It is difficult to estimate the exact number of indris, but their numbers have most certainly been drastically reduced in recent years following the disappearance of their habitat. This species' very slow reproduction rate makes it even more vulnerable, and at present it is very much endangered.

Social Organization

Indris live in family groups formed by a couple with its offspring. These families occupy clearly marked territories, the center of which is defended against intruders. The characteristic plaintive cry, in which all family members participate, serves to delimit the territory and to transmit, up to 5.6 mi (3 km), information relative to the age, sex, and reproductive availability of individuals.

Behavior

The indris' day in these cool altitudes does not begin until two or three hours after sunrise. Feeding accounts for 30 to 60 percent of the activity, peaking at noon. Around 3:00 or 4:00 P.M., the family comes back together to spend the night, and they sleep alone or in twos on the fork of a tree or a large branch.

Reproduction

The female gives birth to a single offspring every two or three years. The infant is carried under the belly until four or five months, then on the back. Maturity is reached between seven and nine years.

Despite its resemblance to a small puma, *Cryptoprocta ferox* is not a cat. With a length of 5 feet (1.5 m), including tail, it is perhaps the largest predator on the island.

Verreaux's Sifaka
(Propithecus verreauxi verreauxi)

FRENCH: sifaka de Verreaux
MALAGASY: sifaka

Description

Average length: adult head and body 16.5–17.7 in (42–45 cm), tail 21.7–23.6 in (55–60 cm).
Average weight: adult 8.3 lb (3.75 kg).

Verreaux's sifaka, with its rather rangy body, includes four subspecies. *Propithecus v. verrauxi*, the subspecies observed in Berenty, is white, with a black muzzle and white forehead, topped by a brownish red crown.

Family

Indriidae.

Locomotion

Thanks to its powerful legs, the sifaka can propel itself by leaps of 33 ft (10 m) from trunk to trunk, landing in a vertical position. Biped on the ground, legs flexed, its ballerina leaps accompanied by acrobatic movements of the arms, give it a rather unique style.

Diet

Although its preferred diet consists of certain buds whose difficult access requires considerable acrobatics, the sifaka mainly eats leaves, fruits, flowers, and the bark of young trees,

Taboos once protected the sifaka from being hunted, but this is no longer the case.

Verreaux's Sifaka

strips of which it carefully lifts with its teeth.

Predators

Raptors, man.

Longevity

The propithecus lives between 25 and 30 years.

Distribution

This subspecies lives in south and southwestern Madagascar.

Subspecies

Formerly vast, the sifaka's habitat has been slowly disappearing, as are the *jady* (taboos) that protected it against hunters.

Social Organization

Troops are mixed, or sometimes consist of single males, with an average of four to six members. These groups are territorial, and they defend a core home range against other groups. Females are dominant. The generally peaceful nature of sifaka changes rapidly during mating season; males confront each other in order to establish the hierarchical rights to mate and to have access to food for the next year.

Behavior

Searching for food represents half of the activity of the sifaka, but playing and grooming are also important. They actively seek the sun on winter mornings, but on summer afternoons they shelter themselves, panting, in the shade of trees.

Reproduction

A single infant is born every two years, generally at the end of July or the beginning of August. The youngster, object of nonstop attention from other members of the troop, remains gripped to its mother's belly for one month, then moves to her back. It is weaned at the beginning of the rainy season, when buds and young leaves proliferate. Maturity is reached at between two and three years.

The young sifaka grips tightly to the belly of its mother during the first month of its life, then it moves to her back.

Ring-tailed Lemur
(Lemur catta)

FRENCH: maki catta
MALAGASY: maki or hira

Description

Average length: adult head and body 16.7 in (42.5 cm), tail 24 in (60 cm).
Average weight: adult 7–7.7 lb (3–3.5 kg).

The best-known coat of the lemurs is pearl gray, with a whitish belly. The face is white and the nose is black. The eyes are surrounded by dark triangular patches. Its most recognizable feature is perhaps its long, white tail ringed with black.

Family

Lemuridae.

Locomotion

This, the most terrestrial of the lemurs, spends up to 40 percent of its time on the ground. During travel, the members of a troop never lose sight of each other, keeping their tails lifted up in the air.

Diet

The ring-tailed lemur feeds on leaves, flowers, and fruits from a wide variety of plants, but it sometimes also eats insects and small vertebrates. It needs a great deal of water that can be found in the hollows of trees, where it inserts a hand, then licks the water off.

A very playful animal, the ring-tailed lemur lives in groups of 3 to 25 individuals.

Predators

Raptors, man.

Longevity

Ring-tailed lemurs live 20 to 25 years.

Distribution

The ring-tailed lemur occupies various habitats in south and southwestern Madagascar; it prefers gallery forests and avoids the moist forests of the east.

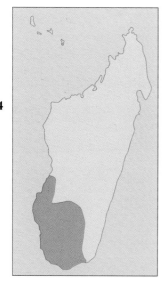

Status

Although it occupies a large number of protected areas, the ring-tailed lemur is hunted over a large portion of its area of distribution. The greatest danger remains the disappearance of its habitat.

Social Organization

Troops consist of 3 to 25 members. Females are dominant and spend their entire lives with their natal group, while males change as soon as they reach sexual maturity. Territory, well marked and stable, is generally defended by vocal contests between families. This is the most vocal of lemurs.

Behavior

Near the wrist the male ring-tailed lemur possesses an odoriferous gland, overlaid with a horny spur. It marks its territory with the former, after having delimited it with the latter. When annoyed, the male rubs his tail between his wrists. Saturating it with his odor, he hits it against his head, projecting his odor 10 ft (3 m) around him in the direction of a potential rival. The latter runs away, growling, before returning to project his own odor; thus, real battles can be avoided.

These characteristics, both morphological and behavioral, are unique, and explain why the ring-tailed lemur is increasingly considered to be a separate genus.

Reproduction

A single infant, more rarely twins, is born in August or September. Literally attached to the belly of its mother, the baby begins to move to her back at around two weeks, before moving off one or two weeks later to play, like a kitten, with other youngsters. Weaning takes place at five or six months. Females give birth for the first time at three years of age.

Lesser Bamboo Lemur
(Hapalemur griseus)

FRENCH: petit hapalémur
MALAGASY: bokombolo, kottika

Description

Average length: adult head and body 11 in (28 cm), tail 14.5 in (37 cm).
Average weight: adult 25–35 oz (700–1,000 g).

The coat of the lesser bamboo lemur is gray, the head round and reddish, the face rather flat, and the small ears covered with hair.

Family

Lemuridae.

Locomotion

This quadruped runs as quickly on the ground as it does in the trees, and is capable of executing great leaps.

Diet

Feeding almost exclusively on bamboo, it prefers the shoots and the bases of leaves but also chews on young branches in order to get to the tender, edible parts. Many of the lemur's teeth are serrated to help process this diet. When the bamboo is too ripe, it eats palm fruit, as well as the leaves of fig trees and various grasses.

Predators

Raptors, man.

Lesser Bamboo Lemur

Longevity

Bamboo lemurs live up to 22 years in captivity.

Distribution

The lesser bamboo lemur is generally found in all of the eastern forests where bamboo grows.

Status

The species is fortunately among the least endangered at the present time; nevertheless, it suffers, as do other lemurs, from the disappearance of its habitat and more widespread hunting.

Social Organization

Mixed troops of four to six members. Females are dominant, to the extent that they have priority access to food and choose their sexual partners. Social bonds are established and maintained by grooming.

Behavior

Although it is the smallest of the diurnal lemurs, its habits tend toward the crepuscular. While secretive, its repertoire of vocalizations is among the most rich.

Reproduction

Ordinarily gentle by nature, the female can become very aggressive after giving birth. A single infant is born per year following a gestation period of 140 days. The first day, the mother carries it in her mouth, then on her back in a transverse position. In captivity, the male sometimes carries it. The mother can also leave her infant alone, sometimes for three hours, in the fork of a tree or other secure place. It waits there perfectly immobile and silent; only the growls of its mother cause it to show itself.

Young lesser bamboo lemurs are very shy and can occasionally be identified by means of the stalks of bamboo that they have stripped of shoots and leaves.

OBSERVING MONKEYS IN ASIA

White-handed or Common Gibbon

(Hylobates lars)

GERMAN: Weisshandgibbon
FRENCH: gibbon à mains blanches
SPANISH: gibón de manos blancas
ITALIAN: gibbone dalle mani blanche

Description

Average length: head plus body between 17 and 25 in (44 and 64 cm).
Average weight: female between 11.6 and 21.1 lb (5.3–9.6 kg); male between 11 and 15 lb (5–7 kg).

The white-handed gibbon is the smallest of the gibbons. As with other gibbons, the arms are long and the legs short. There is no real sexual dichromatism, although females are occasionally somewhat lighter. The thick coat ranges from black to chamois. The white-handed gibbon is easily distinguishable by the ruff of white hair that surrounds its black and naked face, but particularly by its hands and its feet, which are pure white.

Famille

Hylobatidae.

Locomotion

Completely arboreal, gibbons, and their cousins, the siamangs, are the only animals capable of suspending by one hand and pivoting 360 degrees thanks to their extremely flexible shoulders. On the ground, they are also the only primates, outside of man, to move in a truly biped fashion, with feet flat, but, unlike man, they hold their arms at shoulder level or above the head.

Diet

Fruits constitute 70 percent of the diet of this gibbon, the remainder consisting of leaves, flowers, buds, insects, and even eggs. The water they require is generally obtained from fruit, by sucking the rain or the dew off leaves or fingers, or by immersing the arm in a source of water and then licking it off the fur.

Predators

The gibbon moves quickly through the trees, where its only predators are pythons and large raptors. On the other hand, it is vulnerable on the ground, where it can be the victim of land predators such as the leopard.

Longevity

30 to 40 years.

Distribution

The white-handed gibbon is present in Thailand, but also in Burma, within Yunan in China, as well as in Maylasia and northern Sumatra, in Indonesia.

Status

It is in danger of extinction in Burma, China, and Indonesia. The situation in Thailand is less critical, but its commercialization as a household pet remains a real danger.

Social Organization

Monogamous, gibbons live in family groups that typically include an adult male and female, with up to four young. The family occupies a territory that it defends by vocalization duels, resorting to physical violence only as a last resort. The female plays an important part in defen-

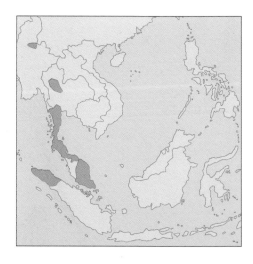

White-handed or Common Gibbon

ding the territory, singing the principal role in the morning duet that announces to neighboring groups the presence of the family. Experiments conducted with the aid of a tape recorder on another species of gibbon (*Hylobates muelleri*) have revealed that it is the female that begins the vocal defense when the previous recording was that of a female, and the male when it was a male. It can thus be deduced that the male and the female drive away their same-sex rivals to safeguard the monogamous relationship.

Behavior

Unlike its close relatives, the great apes, the gibbon does not construct a nest, but sleeps seated on its ischial callosities, knees at chest level, hands on the knees, and head placed between knees and chest. Gibbons are well known for their intelligence.

Reproduction

The menstrual cycle is 28 days. A single infant is born approximately every two years, following a gestation period of 210 days. The newborn is sufficiently developed to grip the fur of its mother as she swings from branch to branch, but it is bare, apart from a crown of hair, and must be sheltered between the thighs and the abdomen of the mother to be warm. The two sexes reach maturity at around seven or eight years of age.

History

Gibbons appear to have diverged from the ancestral stock 16 to 20 million years ago. Fossils dating from the Miocene (18 million years ago) and strongly resembling small gibbons have been found in Kenya; others, somewhat more recent, in Europe. But until the Pleistocene, only a single tooth (from approximately 8 to 10 million years ago), discovered in Asia, could be connected with modern gibbons. However, none of these fossils is conclusive.

Female gibbons actively participate in the defense of their territory.

Observation Site

Thailand: Khao Yai and Khao Phra Thaew National Parks

The Khao Yai National Park houses one of the largest elephant populations in all of Southeast Asia.

Two species of gibbon share the territory of Thailand; the most common is the white-faced gibbon, still called the hylobate. This particularly agile and graceful primate is widely distributed throughout the country, where it occurs in primary or secondary forest areas. The total number of gibbons has declined alarmingly in the last few decades.

We suggest two very different sites for observing gibbons: Khao Yai National Park, where the visitor has a very good chance of observing these graceful apes in their natural environment, and Khao Phra Thaew, on the island of Phuket, where there is a rehabilitation center for these animals. Khao Yai is one of the oldest, the best-known, and most visited of Thailand's national parks; it covers more than 531,050 acres (215,000 ha), and is generally recognized as being one of the most remarkable and well-protected sites in all of Southeast Asia.

Fauna and Flora

Khao Yai National Park covers a hilly region; significant differences in altitude characterize the site, which rises from 820 ft (250 m) to more than 4,428 ft (1,350 m) of altitude, then descend rather abruptly toward the agricultural plains along its borders in the south and east. The park is basically forest, deciduous, dry, or moist. Vast expanses of savannah covered with high grass vegetation, small secondary forests, as well as several swamp areas, complete this diversified ensemble. Epiphytes and orchids, the national symbol of Thailand, flourish in the park, which is also home to more than a dozen species of endemic flora.

Thailand: Khao Yai and Khao Phra Thaew   White-handed or Common Gibbon

Over its vast expanse, Khao Yai National Park includes hilly terrain, basically forest, cooled by several rapid streams.

More than 300 species of birds, including some that are quite spectacular and easily observable, make Khao Yai an important ornithological site in Thailand. The park is home to more than 20 species of large mammals, including the Asian elephant, the guar, the sambar, the barking deer, the Asian black bear, tigers, the leopard cat, and others. Primates are represented by gibbons and macaques.

WHITE-HANDED OR COMMON GIBBON THAILAND: KHAO YAI AND KHAO PHRA THAEW

A fearsome arboreal viper hunting prey in the Khao Yai National Park.

Offense Against Nature

The name gibbon, common in the majority of Western languages, has no equivalent in Asia. It probably originated with the Italian word *gibboso* ("hunchback"), brought back to Europe by Marco Polo. The scientific name, *Hylobates*, means "forest walker."

In China, the gibbon has been considered for more than 2,000 years as the noblest of monkeys. Symbol for poets and philosophers of the ideal of earthly detachment, it also represents the mystery that links man with nature. A fourth-century Chinese historian recounted in a newspaper of the time, that Teng Chih, general of chariots and cavalry, after having repressed the rebellion of the powerful Hsii clan, "...saw a dark-colored gibbon climbing a mountain. Teng, being an amateur archer, shot an arrow at the gibbon, and caught it. The gibbon's offspring pulled out the arrow and tried to stop the flow of blood by covering the wound with leaves. 'Alas,' sighed Teng, 'I have committed an offense against Nature'."

Others freely shoot at mothers to take their young—a technique that is still widespread today, particularly in Thailand—to make pets of them. Debates surrounding captivity are as old as the tradition itself. Tung-po (1036–1101) wrote: "Once a monkey becomes used to human clothing, if you watch it carefully, you will note boundless contempt. People claim they play with monkeys, but I wonder if it isn't rather the monkeys that are playing with men."

Observation

The open expanses that are scattered throughout the Khao Yai forest are exceptional sites for observing fauna and avifauna. The observation of gibbons, spread widely throughout the park, is relatively easy, particularly near the headquarters, where the visitors center is also located. The best times of day are early morning and the end of the day; it is at both these times that it is possible to hear the resonant, melodious cries emitted by the gibbons to define their territory.

One of the best spots for observing these primates is the path that leaves the visitors' center, the King Kaew Falls Nature Trail. This trail is also good for ornithological observation.

The opportunities for observing gibbons in the Khao Phra Thaew National Park are totally different. This small national park, approximately 8,200 acres (3,320 ha) in size, protects the last expanses of primary tropical forest that have survived on the island of Phuket, densely populated and urbanized, which has become one of the most fashionable tourist sites in Thailand. It is in this protected site, with its hilly terrain, that the Rehabilitation Center was established. This center, directed by a nongovernmental association, is an initiative of the American biologist Terence Dillon Morin, who became interested in the unhappy situation of gibbons in captivity that he discovered on the island. The purpose of the center is to recreate families of gibbons, and then to progressively rehabituate them to life in the forest, in order to repopulate certain sites where these animals have been exterminated. Several groups have already been returned to the wild on an island of 2,470 acres (1,000 ha) of forest located in Phangnga Bay. The Rehabilitation Center is located near Bang Paie Waterfall. It is possible to take a guided tour of the facilities, which allows one to observe the gibbons and to learn more about their behavior and the problems of their conservation in Thailand.

PRACTICAL INFORMATION

Our winter is the best time to visit the Khao Yai park.

Khao Yai National Park

TRANSPORTATION

Many regularly scheduled flights link Bangkok with a majority of capitals.
■ **BY CAR.** Khai Yai National Park is located approximately 99 mi (160 km) north of Bangkok. Numerous tourist agencies organize excursions to the park from the capital. It is also possible to rent a car or to hire a taxi to reach the park from Bangkok along excellent asphalt roads. Various bus services (regular or deluxe) also leave Bangkok for the park each day.

ACCOMMODATIONS

Various types of accommodations exist in Khao Yai: from the rather luxurious and relatively expensive lodge to camping sites and bungalows. All of these establishments are located in the immediate proximity of the park headquarters and visitors' center, located near the main entrance. It is also possible to find lodging in the area of Pak Chong, located near the main entrance to the national park, where various hotels, motels, and guesthouses are available.

CLIMATE

In Khao Yai, the climate is determined by the rhythm of the monsoons—from July to October, the rains are abundant, while the dry season begins in November, when there is much less precipitation. Various areas of the park sometimes experience significant local variations, particularly determined by the altitude and the exposure of the slopes. The highest temperatures—up to 86°F (30°C) are recorded from March to May; in December and January, the temperature often falls below 50°F (10°C) in the evening.

TRAVEL CONDITIONS

The Khao Yai National Park charges an admission fee; a permit can be easily obtained at the visitors' center. The park is open all year. It is possible to hike the trails independently, but the services of a guide are recommended, if only for their ability to locate the animals, which is invariably better than those of visitors.

Khao Phra Thaew National Park

TRANSPORTATION

■ **BY PLANE.** Thai Airways departs daily from Los Angeles (you can connect from San Francisco, Seattle, Dallas, and New York). Phuket is accessible directly from overseas—there are numerous charter flights, as well as several regular airlines departing from Europe that land directly in Phuket—or by car, train, or plane from Bangkok.
■ **BY CAR.** Khao Phra Thaew is located in the center of the island, east of the main road that traverses Phuket from north to south. It is not easy to get there by public transportation, but it is easy to rent a taxi, a motorbike, or a car to get there, the distances being rather short, approximately 12 mi (20 km).

ACCOMMODATIONS

It is not possible to stay in the immediate vicinity of Bang Pae Falls, or even to camp there, but tourist establishments (hotels in all categories) abound on the island of Phuket, principally along the west coast.

CLIMATE

On the island of Phuket, the rains have a tendency to fall almost all year, with the maximum rainfall occurring in May and in October. Precipitation often occurs in the form of violent tropical storms, particularly frequent at the end of the afternoon and in the evening. Temperatures are rather constant all year, generally around 77–86°F (25–30°C) in the daytime, and not much lower at night.

TRAVEL CONDITIONS

A visit to the park is subject to a modest entry fee; it is important to note, however, that there is no real trail or path in the forest so a guide is indispensable. The number of tourists hiking in Khao Phra Thaew being rather low, despite the fact that more than a million tourists land on the island each year, it is often preferable to indicate your intention to hire a guide one or two days in advance to the park administration, which maintains an office near the Gibbon Rehabilitation Center. It is also possible to participate in trips organized by several agencies from Phuket Town into the forest.

Japanese Macaque
(Macaca fuscata)

GERMAN: Rotgesichtmakak
FRENCH: macaque du Japon
SPANISH: macaco japonés
ITALIAN: macaco del viso rosso

Description

Average length: male head and body between 19.7 and 26 in (50–65 cm), tail 2.8–3.5 in (7–9 cm); female head and body between 18.5 and 23.6 in (47–60 cm), tail 2.8–3.5 in (7–9 cm).
Average weight: male between 19.8 and 26 lb (9–12 kg); female 10 lb (4.6 kg); newborn 18 oz (500 g).
Like all animals in regions with cold climates, the morphology of the Japanese macaque is adapted to avoid the loss of heat: The body is stocky and the limbs are short; the small ears are hidden in long, dense fur. The populations in the north even have an undercoat.

Family

Cercopithecidae.

Locomotion

Both terrestrial and arboreal, Japanese macaques occasionally move on two legs (see following page). While the majority of monkeys are capable of swimming, if need be, some better than others, only the Japanese macaque appears to swim for pleasure.

Diet

In winter, the diet of certain populations is limited to tree bark and shoots from the ends of branches, but in the spring, young leaves and fruits, arthropods, crayfish, and birds' eggs are added. Japanese macaques are also fond of wild mushrooms. Such liberal appetites lead the Japanese macaques to accept human handouts and also to raid crops.

Predator

Traditionally protected, despite limited hunting for its meat, its fur, and for pharmaceutical reasons, today, farmers represent the greatest danger.

Longevity

More than 30 years.

Distribution

Living on the high plateaus and coasts of Japan, this macaque is the most northern of primates, outside of man.

Status

The Japanese macaque is still common but localized, and deforestation in favor of agriculture is putting it in a position of serious competition with man. In fact, faced with the plundering of their fields, in 1997 peasants launched an extermination campaign, killing several thousand animals. Thus, free-living populations are in a precarious state.

Social Organization

Japanese macaques live in multimale groups of 20 to 100 individuals. The strong bonds between females assure the cohesion of the troop. Young females remain there in the strictest hierarchical order, the high-ranked maternal lines prohibiting the social ascension of their inferiors. Young males form bands on the periphery of the troop. The male adult that shakes tree branches with greatest energy and noise wins the largest number of matings.

JAPANESE MACAQUE

Behavior

The behavior of this monkey is as exceptional as its physique. While the populations of the snowy valley of Jigokudani slide into the warm water of thermal springs, where the temperature can reach 104°F (40°C), their relatives on the island of Koshima have given astounded scientists an amazing demonstration of behavioral evolution. It all began in 1965 when a young female, Imo, discovered that she could remove the sand from her potato by dipping it into the still water. It didn't take long for her play companions, then their mothers, to imitate her, and before long, all of them were dipping their potatoes in seawater after each bite, this time to salt them. Later, Imo, a kind of genius among monkeys, made a new discovery: She noticed that by throwing grains of wheat into the water, the sand that was mixed in with them fell to the bottom, allowing the cleaned grains to rise to the top. Once again, the others followed her example. New behaviors of this type generated others; as a result of carrying the food to the water, the monkeys became accustomed to walking upright and developed a taste for swimming, and even diving. Today, these learned behaviors are transmitted from generation to generation, something that was believed impossible not so long ago.

Reproduction

Among populations in milder climates, the mating season is not distinct, and couplings take place throughout the year, even outside of the periods of heat. On the other hand, populations in cold areas limit their mating to the months from November to March, in order to assure the birth of the young during the warm months from April to August. The absence in females of edema of the sexual skin during heat is compensated for by the intensity of its redness, which is also found on the face.

History

What is fascinating in this species is not so much its past history as its flagrant evolution, which is taking place before our eyes.

In this amazing species, new behavior is transmitted first between companions, then to mothers.

Observation Site

Japan (Honshu): Jigokudani Onsen

Japanese macaques are remarkable animals with numerous expressions, and their characteristic behavior, as well as their amazing capacities for adaptation and learning, have made them the object of intensive study on the part of scientists at a least one dozen sites in Japan, and even to become famous worldwide.

One of the most surprising behaviors among macaques is the taking of hot baths to fight against the rigors of winter. This exceptional behavior is the prerogative of a single population of macaques; they live on the island of Honshu, one of the principal islands of the archipelago, under latitudes that make these animals the northernmost monkeys in the world. Winter, which is harsh in the center of the island, is the season of trials and privations for the macaques.

Jigokudani Valley (Hell Valley in Japanese), where the macaque population in question lives, has been known for ages by the inhabitants of the region for its numerous hot springs of volcanic origin (*Jigokudani Onsen*). It is located in the central part of the island of Honshu and extends into the Shiga mountain range, in the prefecture of Nagano. Traditionally, the Japanese have profited from the benefits of these waters by taking long baths, first most likely for religious reasons, and more recently for curative purposes.

For as long as can be remembered, the monkeys have always lived in this valley, but it was necessary to wait until the 1960s before one of them—a female that is now a matriarch named Tokiwa by the scientists studying the animals in this location—had the idea of imitating humans and entered the water. The experience must have charmed her, because this female, aged three at the time, made it a habit, which was rapidly imitated by the other members of the troop. In only a few years, the winter baths in hot springs had become one of the principal behavioral traits of the entire macaque population of Jigokudani, and this is still the case at present.

In such behaviors, we see the beginnings of what might be called culture, wherein information is transmitted by behavioral means. This differs greatly from genetic transmission, which occurs only between parent and offspring.

Observation

Winter observation of Japanese macaques in the Jigokudani Valley is exceptional, and attracts an increasing number of visitors every year. At the beginning the monkeys became accustomed to bathing in the natural pool near the only hotel

Wise Monkeys

It is not surprising that the three principles of Tendai Buddhism—see no evil, hear no evil, speak no evil—are personified by Japanese macaques. A comparative study was made, not on the monkeys themselves, but on the ethnologists who were observing them. It seems that while Japanese researchers emphasize the importance of sharing resources and cooperation for the survival of the monkeys, the work of Western researchers emphasizes the law of the strongest.

This opposition reflects both cultural and philosophical differences. The Japanese assume quite naturally that the monkey has a "soul" and a personality. The use, with regard to monkeys, of the word *san*, which is a form of address normally used for humans, is proof. Even nonbelievers attend the memorial services in the Osaka temple that houses the tombs of some 20,000 monkeys. In the West, on the other hand, the existence of an animal soul has been the subject of debate since the "soul of beasts" of Antiquity; thus we can see how scientific objectivity can be colored by culture. Perhaps we should compare notes in order to arrive at a more global understanding of life. What can these wise monkeys still teach us?

See no evil, hear no evil, speak no evil. Monkeys have often symbolized wisdom.

The behavioral evolution evidenced by macaques is not the least of their particularities. The habit developed by the population of Jigokudani since the 1960s of bathing in the warm water of thermal sources has made them world famous.

JAPANESE MACAQUE

JAPAN (HONSHU): JIGOKUDANI ONSEN

Despite its name of "Valley of Hell," Jigokudani has a climate comparable to that of the Alps.

in the valley, but it was soon necessary to build them a bathing pool of their own. This pool today constitutes the principal center of attraction for visitors who have come to view the monkeys. Some of the monkeys still continue to bathe occasionally in the hotel pool, even if it is already occupied by humans, which gives rise to a cohabitation that is as peaceful as it is incredible.

The macaques are present at the site all year, and can, therefore, be observed during any season, but they bathe in the warm springs only during the winter. Outside of this period, they have a tendency to spread out in the pine forest that covers a good portion of the Jigokudani Valley, which can occasionally make observation rather uncertain.

PRACTICAL INFORMATION

TRANSPORTATION

Many regularly scheduled flights make it possible to reach Tokyo from major cities throughout the world.

■ **BY TRAIN.** Various means of transportation lead to the city of Yudanaka, the closest to Jigokudani Valley: train, bus, or car. The train is perhaps the most comfortable option; by car, it takes approximately eight hours departing from Tokyo.

■ **BY CAR.** The valley is located some distance from the city of Yudanaka so it is better to have your own means of transportation to reach it, or at least to use the relatively expensive services of a taxi, which should be reserved for the return trip.

■ **ON FOOT.** The hot springs (onsen in Japanese) are approximately 1.2 mi (2 km) from any road, a distance that must be covered on foot on a mountain path.

DISTANCES

Tokyo to Yudanaka: 435 mi (700 km)

ACCOMMODATIONS

Accommodation possibilities are numerous in Yudanaka, but it is better to find something in the immediate area of the springs in order to maximize the possibilities for observing macaques. A single rustic hotel (Koraku-Kan) exists in the valley. Constructed a long time ago on a promontory, it dominates the hot springs and makes it possible to reach the "monkey springs" in only ten minutes of hiking. The capacity of this establishment is limited, and, considering the popularity of the site with visitors in winter, it is often necessary to reserve long in advance.

CLIMATE

The climate of the Jigokudani area is temperate, comparable to that of the Alps in Europe. In winter, the period of hot baths for the monkeys, snowfall is often significant, and freezing temperatures can last for several weeks at a time.

TRAVEL CONDITIONS

No permit is required to view the monkeys in Jigokudani; however, it is not possible, for obvious reasons, to enter the water in the springs where these animals bathe, or to get too close to them without specific authorization.

It should be noted that macaques have become rather reckless and impudent, and they do not hesitate to attack individuals visibly transporting food—they seem to fear only the forest rangers and the hotel employees. Therefore, every precaution should be taken not to appear to have anything resembling food, especially between the hotel and the hot springs, in order to avoid any accidents. It is, of course, prohibited to feed the macaques; they receive an appropriate diet of grains of wheat and such from the forest personnel when the weather conditions become too harsh.

Celebes Black Ape
(Macaca nigra)

GERMAN: Schopfmakak
FRENCH: macaque noir de Célèbes
SPANSIH: macaco negro
ITALIAN: macaco di Celebes

Description

Average length: male head and body between 20 and 22 in (500 and 570 mm), tail between about ½ and ¾ in (15 and 24 mm); female head and body between 17.5 and 21.7 in (445 and 550 mm), tail 0.6 in (16 mm).

Average weight: unknown. Males are completely black, females generally a bit lighter. The long face with the protruding occiput (back of the head) is reminiscent of that of baboons. The ears are bare or sprinkled with white hairs. The tail is reduced to a simple stump. The adult celebes ape is distinguished by the conical crest of hair on the crown.

Family

Cercopithecidae.

Locomotion

The celebes ape lives in the forest, where it moves on four legs both in the trees and on the ground.

Diet

It basically feeds on fruits, flowers, young leaves, shoots, insects, and caterpillars.

Predators

Large snakes, birds of prey, crocodiles, and man.

Longevity

Among all the macaques, the celebes ape holds the record for longevity. In captivity, one of them lived for 28 years.

With its long face, the black macaque resembles a baboon, but it is much closer to the pig-tailed macaque.

Celebes Black Ape

Distribution

The species is present in the extreme northeast of the Celebes as well as on the islands of Lembeh, Manadotua, and Talise.

Status

This is an endangered species. For several generations, the population has decreased by half, especially due to the reduction and the impoverishment of its habitat.

Social Organization

The celebes ape lives in small multimale troops. These troops range in size from five to twenty-five individuals and are generally led by one old male. It does not have the white eyelids that make it possible for many other monkeys to express their sentiments. On the other hand, it uses a seemingly shrewd gesture as an invitation or when approaching others—it lifts the eyebrows while lowering the ears, which inclines its crest toward the back. Also as a sign of friendship, but common among a large number of monkeys, it smacks its lips, sometimes quickly with the tongue. This smacking of the lips may be exaggerated in an aggressive animal, which indicates a conflict between the desire to attack and the desire for peace; however, an aggressive celebes ape is generally recognized by its rigid gait and its fixed look.

Self-grooming along with the grooming of others is frequent. It often takes place in a chain formed by several individuals. An animal looking to be groomed often assumes one of the following two postures: It stretches out on its side, back turned in the direction of the animal whose attention it wishes to attract with all four limbs extended to the maximum, or it stands up, both legs rigid but arms folded in such a way that the head is completely bowed and the thighs are out in the open. Other species may do the same once grooming has begun, but only the celebes ape is this forward. Hierarchy plays very little role in grooming relationships.

Behavior

In analyzing the reaction of different species of celebes apes (whose number varies according to the authors) to the projection of slides, it was determined that they know how to distinguish their own species from others, but also similar species from dissimilar ones. Males manage quite well at this task, with the sole exception of the celebes black ape, with the female not doing much better. This recognition or absence of recognition may have an influence on the already high rate of hybridization in this species, which explains, in part, the disputes among primatologists with regard to their exact number.

Reproduction

Little is known about this monkey, but it is known that during heat, the edema of the sexual skin appears in the female. A single infant is born following a gestation period of 162 to 186 days.

History

The celebes apes are among the rare species of monkeys that left Asia to cross Wallace's line, a hypothetical border imagined in the nineteenth century by the scientist of the same name, and that separates the fauna from Southeast Asia from that of the southern region. It is currently thought that all these macaques are derived from a single species that may have strongly resembled the pigtail macaque. The celebes ape would be the species furthest removed from this ancestor.

Observation Site

Indonesia (Sulawesi): Tangkoko Duasudara Natural Reserve

The Tangkoko Duasudara Natural Reserve—often incorrectly called the Tangkoko Batuangus Natural Reserve, the name of the larger forest reserve in which it is located—is situated in the extreme northeast of Sulawesi (or Celebes). It covers almost 22,230 acres (9,000 ha) of forest that reach as far as the eastern coast of the island.

Created by the Dutch colonial authorities around 1919, it protects several thousand acres of moist tropical forest of the plains, as well as several expanses of mangrove and a coral reef located a short distance from the coast. It is home to a remarkable number of animal species, which motivated the designation of the site as a reserve. Tourism developed there only recently, since the end of the 1980s. It is based on the observation of primates and birds.

Fauna and Flora

The natural environment of the reserve is relatively uniform, consisting primarily of moist tropical primary forest; along the coast, however, the presence of mangroves and marine environments brings noticeable variation to the countryside.

The most remarkable species in Tangkoko Duasudara are the primates; the celebes black ape, still called the crested macaque, is in fact common throughout the forest, as is the tarsier. Hornbills, of which several species exist in the reserve, are also one of the principal attractions of the reserve, which is also home to five endemic species of kingfishers.

The babirusa still survives there but has not been officially seen for several years; the anoa (the world's smallest cattle species) and the maleo (the only bird that does not sit on its eggs) have also become quite rare.

Observation

Observation conditions for celebes apes and tarsiers can be called exceptional in Tangkoko Duasudara: The large majority of visitors making a trip into the forest, in fact, have the opportunity to observe these

Timid, shy, and increasingly rare, the babirusa is among the strangest and most mysterious inhabitants of Tangkoko.

animals that are generally not noticed and timid elsewhere.

Three groups of apes, each numbering between 50 and 100 animals, are used to visitors, while several daytime dormitories of tarsiers are known by the guides in the reserve, which makes it possible to observe these basically nocturnal animals.

Tourism in the Tangkoko Duasudara Natural Reserve has recently been subject to severe criticism, and a study commissioned by the authorities has just been conducted in order to evaluate the actual impact of visits on the animals and the natural environment. The status of natural reserve excludes tourism based on Indonesian law concerning nature conservation. Tangkoko Duasudara is an exception to this rule, which has both positive and negative consequences. The resources generated from tourism to observe the animals have, in fact, the potential of increasing the protection of the site and the forest reserve that surrounds it, making the inhabitants conscious of the economic importance the fauna represent. On the other hand, it appears that the managers of the reserve have allowed tourism to develop in a somewhat anarchic manner, which has harmful consequences for the behavior of the animals most sought after by visitors, such as macaques and tarsiers. The reserve may next be elevated to the status of national park, which would make it possible to manage tourism in a more efficient manner, based on

Man's hold over nature in Southeast Asia, a region often overpopulated, constitutes one of the greatest dangers for primates, whose habitat is being slowly reduced (seen here, a vast expanse of rice paddies around a village in the Celebes).

the legal requirements of the regulations governing this activity in national parks. In the meantime, it is especially important that visitors behave respectfully toward the animals by not getting too close, avoiding any behavior capable of frightening the apes, or disturbing the sleep of the tarsiers, and their guides are required to do the same. Perhaps more than elsewhere, the long-term survival of the primates of Tangkoko Duasudara—and of tourism, itself—depends on it.

> **Crocodile Tears**
>
> In the tradition of the Celebes, the celebes ape is often accused of great impudence, but its pranks always indicate a great deal of intelligence and wisdom. Its principal enemy is the crocodile, and numerous stories recount how, even in extreme situations, the ape wins the day. This is the case in the story where the crocodile catches a monkey by the leg and the latter, by a series of ingenious arguments, manages to convince the crocodile that it has bitten off a piece of wood by mistake. Finally, the reptile opens its mouth to see, and the monkey takes advantage of the situation to escape.
>
> In another legend, a crocodile locks up an ape on a small island with the intention of eating it there. The ape convinces the crocodile that the feast would be more fun if other crocodiles took part. After serious reflection, the crocodile calls a dozen of its companions. The ape, pretending that its poor eyesight prevents it from seeing the magnificent saurians that are about to eat it, asks them to line up in single file in order to be able to count them more easily. The crocodiles line up and the ape leaps from one to another with lightning speed, to reach land, safe and sound.
>
> It's enough to make a crocodile cry.

PRACTICAL INFORMATION

Driving conditions sometimes make it impossible to travel during rainy season.

TRANSPORTATION

Sulawesi can be reached via Jakarta. Daily flights are scheduled by American Airlines, Singapore Airlines, and Japan Airlines from Los Angeles and New York, and by British Airways from New York. The voyage is continued on an internal flight as far as Ujungpandang. Private flights, whose frequency is a function of demand, link the towns of Ujungpandang and Manado.

■ **BY CAR.** Access to the reserve is easy, since a road makes it possible to cover the distance between Manado, the principal town in the area, and the reserve (31 mi, 50 km) in approximately two hours. The reserve is also 18.6 mi (30 km) from the port of Bitung.

While there are several villages in the immediate area, it is Batuputih that has become the principal point of entry into the reserve. This village is located near the main road from Manado.

■ **BY BUS.** Bus service provides a link from Manado, but it is not always reliable and is rather uncomfortable. It is possible to rent a vehicle in Manado, and several tourist agencies organize trips to the reserve departing from this location.

ACCOMMODATIONS

No accommodation infrastructure exists in the reserve. Three rustic guesthouses, with rather basic comforts, make it possible to stay in the village of Batuputih. No other possibility for lodging is available at the moment, outside of the city of Manado itself, where the options are more numerous.

CLIMATE

The climate in Tangkoko Duasudara is markedly similar throughout the year—invariably hot and humid. The rainy season, in which the precipitation is most intense, extends from November–December to March–April; however, there are storms in all seasons.

TRAVEL CONDITIONS

A visitors' permit must be obtained in advance; it must be requested at the PHPA (Indonesian National Parks Administration) in Manado.

Guides are available at the entry to the reserve as well as in the village of Batuputih; they are generally guards in the reserve, and occasionally villagers when the guards are not available. There is a fee for the services of these guides, and it is not permitted to enter the reserve without being accompanied by an individual officially authorized to guide visitors.

Orangutan
(Pongo pygmaeus)

GERMAN: Orangutan
FRENCH: orang-outan
LOCAL: Orang-hutàn

Description

Average length: male head plus body 3.1 ft (95 cm); female head plus body 2.5 ft (77.5 cm).
Average weight: male 169 lb (77 kg); female 81 lb (37 kg).

After the gorilla, the orangutan is the largest of the primates. Its body is covered with long red hair; the thumbs of its long hands, like those of its feet, are reduced. The male is two times heavier than the female.

There are two subspecies; that of Sumatra has a narrower face and generally lighter hair. The male often has a long beard and mustache that give him the look of a wise old man. This is not the case with the males in the Borneo subspecies that have large cheek pads.

Family

Pongidae.

Locomotion

Orangutans are almost completely arboreal, and occupy all of the levels of the forest. The large size of the adult prevents the orangutan from moving by true brachiation; it thus progresses by distributing its weight equally on all four limbs and often by testing a branch before grasping it. This prudence increases with age and the weight that accompanies it. Smaller animals can swing from branch to branch. They descend to the ground only on rare occasions.

Diet

The orangutan is the largest of all the frugivores. It basically spends most of its time meeting its need for food. To this end, it consumes more than 300 different types of food. From April to November, its diet consists primarily of fruits (durians, plums, and wild figs, bread tree fruit, rambutans, and others), which it consumes in large quantities in order to build up fat stores. During the monsoons, it eats leaves, bark, and bamboo. It will

The orangutan can, on occasion, cross a stream.

Orangutan

occasionally eat ants, termites, eggs, small vertebrates, and bees.

Predators

In areas where there are still tigers and leopards, females may spend 95 percent of their time in the trees. As for the majority of primates, the greatest danger comes from man, either directly by poaching for food or commercial supply to zoos and pet stores, or indirectly by deforestation.

Longevity

Approximately 50 years, the record being 59.

Distribution

In Borneo, orangutans live in the tropical rain forests of the plains. In Sumatra, they are limited to a mountainous area.

Status

In 1996 the number of orangutans was estimated at less than 30,000, which already represents a decline of from 30 to 50 percent in ten years. Since the ravaging fires of 1997–1998, this number has fallen again. Today, it is difficult to determine the exact number of animals lost, but according to some authorities, these fires may have pushed the total number below the precarious threshold required for the survival of the species.

Social Organization

The orangutan is the most solitary of the primates. The male occupies a large territory that covers the smaller territories of several females, constituting his harem. They themselves live with only one or two youngsters maximum, perhaps because of the large quantity of food that these large animals require and the fierce competition that results. The howling of the male can reach 3/5 of a mile (1 km) and serves to warn possible intruders. The ranges of females within an area generally overlap, and loose aggregations of several animals have been reported.

Behavior

The orangutan is exceptionally intelligent, even more so than the chimpanzee, according to some scientists. In the evening, it sleeps in the trees at 19.7–77 ft (6–24 m) of height, in nests constructed from leaves.

Reproduction

The menstrual cycle is 29 to 30 days, during which the female remains sexually receptive. There is no sexual edema during estrus, but sometimes it occurs in the external genitalia during gestation, which lasts 264 days. The young are weaned at three years, but the mother continues to carry them for another year. The latter is extremely indulgent toward the mischievousness of her offspring. Maturity is reached at six or seven years for females, at ten for males. A female has only two or three offspring during her lifetime, which makes the current threats even more disturbing.

History

Orangutan teeth have been discovered in caves in southern and central China, indicating their long presence on the continent. The thick enamel of their molars leads one to think that their ancestors were more terrestrial, terrestrial monkeys having a diet that is harder and more fibrous than that of arboreal monkeys.

Observation Site

Indonesia (Sumatra): Bohorok Rehabilitation Center, Gunung Leuser National Park

The illegal trade in baby orangutans continues despite laws designed to protect the species (shown here, a newborn rescued by forest guards in Borneo following a fire).

Indonesia is one of the countries in Southeast Asia that has the most primates, among which the orangutan is one of the best known and the most remarkable.

Formerly present throughout a wide area of distribution, the "person of the forest" (translation of the Indonesian term *orang-hutàn*) survives today only in Sumatra and Borneo. Almost everywhere, its numbers have unfortunately been reduced by uncontrolled and illegal hunting and the destruction of its habitat—the tropical forest—particularly following fires, such as those that ravaged several million acres in 1997.

Gunung Leuser (Mount Leuser) National Park is today one of the most important sanctuaries for this endangered species. Located in western Sumatra, some 62 mi (100 km) from the city of Medan, it owes its name to the second highest peak in Indonesia, Mount Leuser, which towers over it at 11,300 ft (3,445 m). The park covers more than a million acres, a good portion of them on the mountainous slopes of Barisan, which crosses the island from one end to another. Three somewhat parallel mountain formations follow it, cut by wide valleys and several rather deep canyons.

Created at the end of the 1970s, the park is divided into four regions, each benefitting from different management measures, particularly with regard to the infrastructures for visits.

Fauna and Flora

The large area of the park, as well as the differences in altitude, explain the diversity of the natural environments found there, ranging from tropical primary forest, which is most of the park, to secondary as far as Alpine-type plains, and including swampy forests.

Gunung Leuser is home to remarkable and often rare species, such as the Asian elephant, the Sumatra rhinoceros—almost 20 percent of the total number of this very endangered species are found in the park—the Sumatra tiger, the Malaysian sun bear,

The mist often dominates in the high forest of tropical Asia, a natural environment saturated with moisture.

Orangutan

Indonesia (Sumatra): Gunung Leuser

Young orangutans in a "family" in Borneo. As soon as they reach the age of maturity, these large primates lead a basically solitary existence.

the cyon (wild dog), various species of *Felidae*, including the leopard cat, the golden cat, and the leopard, along with several species of deer, including the sambar. No less than seven species of primate also live there, including the long-tailed macaque, the most frequently sighted. In total, Gunung Leuser is home to some 320 species of birds, 176 species of mammals, 194 species of reptiles, and 52 species of amphibians.

Observation

The closed environment of the tropical forest in Sumatra often makes it difficult to directly observe animals, even the largest of them, and requires a certain amount of patience,

Linnaeus' Reasoning?

Some animals could not be classified by scientists until progress in maritime transportation made it possible for them to have a sufficient number of individuals of each species. For almost two centuries, a certain confusion reigned with regard to the great apes. The scientific name of the orangutan, *Pongo*, was originally the indigenous name for the gorilla. At the same time, the orangutan was called "chimpanzee." The confusion was not only limited to great apes, but also to the distinction between man and the great apes. The indigenous words *orang-hutàn*, in fact, means "person of the forest" (Anthony Burgess, author of the celebrated novel *A Clockwork Orange*, on which Stanley Kubrick based his well-known film, lived in Maylasia for six years, and confesses to having used a play on the indigenous word *orang*, "man," for his title). The term was used in the eighteenth century without understanding why by the father of modern taxonomy, Carl Linnaeus. The latter classified the orangutan as a species of man under the name of *Homo silvestris* ("person of the forest" in Latin). Linnaeus confessed that he had not been able to "take from the principles of his science any characteristic that could be used to make it possible to distinguish the person of the forest." For many, the orangutan represented a silent link between man and the lesser animals. The British scientist, Lord Monboddo, affirmed in no uncertain terms that the orangutan was a man, but a man without language. Descartes, supported by the naturalist Buffon, had, however, claimed that, without speech there is no thought, and thought alone is the distinctive feature of man;

It is not clear what period this engraving is from or what species it was supposed to illustrate. Officially, an orangutan.

this thesis predominated until the 1960s. Since then, progress in neuropsychology has shown the linguistic capacity of great apes, while progress in genetics has reduced even more the distance that was believed to separate us from these close relatives. Perhaps Linnaeus was not quite so mistaken.

and even philosophy, on the part of the visitor. Walking through this very unique forest environment is certainly a memorable experience, with the observation of animals considered a plus and not a formal guarantee.

Fortunately, the observation of orangutans is the exception to this rule. Two rehabilitation stations for these primates have been established in the park. The first is located at the Ketambe Scientific Research Center; however, as this institution is exclusively devoted to research, access is not authorized. The second station, Bohorok, is better known; it is easy to observe orangutans there, either under the rather more artificial conditions of the rehabilitation station itself, or more naturally, in the forest, where several dozen individuals have been returned to the wild, and have thus lost most of their natural fear of man.

The best time of day to observe orangutans in the Bohorok area is at the end of the afternoon, as this is the time when the apes returned to the wild but still not capable of totally fending for themselves return to the feeding station located near the center, where food is made available to them. At other times of the day, the center itself can be visited, or you can take your chances in the adjacent forest area.

PRACTICAL INFORMATION

Do not forget to bring repellents for mosquitoes and leeches, which are rather numerous throughout the forest (except in high altitudes).

TRANSPORTATION

Sumatra is reached via Jakarta with daily flights scheduled by American Airlines, Japan Airlines, and Singapore Airlines from Los Angeles and New York, and by British Airways from New York, and by continuing the trip via internal flight as far as Medan.

■ **BY BUS OR BY CAR.** An asphalt road links Medan to the village of Bukit Lawang, located approximately 62 mi (100 km) from each other, the principal point of entry into the national park. Various bus lines provide the link. The trip can also be made in taxi—the fares are approximately 15 to 20 times that of the bus, not excessive for the distance if several passengers share the cost of the trip—or by renting a car, the most expensive alternative. The taxi and the car have the advantage of making a vehicle available to easily reach the Bohorok area, at the entry to the national park, which is located several miles from Bukit Lawang.

■ **ON AN ORGANIZED TOUR.** The visitor with limited time, or who prefers not to go to Gunung Leuser on his or her own, can join an organized tour leaving from Medan.

ACCOMMODATIONS

Several modest establishments make it possible to stay in the village of Bukit Lawang; in the park itself, the only option is camping in the wild, except in Bohorok, where a campground equipped with sanitation facilities, along with several pleasant bungalows, is available. Food must be provided, along with all of the equipment required for camping, preparing meals, and sleeping.

CLIMATE

The climate at Gunung Leuser is tropical humid; temperatures experience very little seasonal variation, remaining around 86°F (30°C) all year. Of course, the temperature decreases as the altitude climbs, and the nights are freezing in the vicinity of Mount Leuser.

It rains all year in this area of Sumatra due to the influence of the high mountains, but the rains are particularly intense during the humid season, from October to April. The humidity in the forest is very high in all seasons, and is often oppressive. A thick fog often hides the view at high altitudes, especially in the morning.

TRAVEL CONDITIONS

Every visitor must obtain a visitor's permit in advance from the PHPA (Indonesian National Parks Administration) located near the village of Bukit Lawang.

An official guide is required for a visit to the park. The Bohorok area is the most developed from the point of view of tourism, and has many pedestrian paths making it possible to walk, and even to hike. For longer hikes of several days, it is better to use the services of porters to transport food and equipment.

Hanuman Langur

(Presbytis entellus)

GERMAN: Hanuman Langur
FRENCH: entelle
SPANISH: Langur hanuman
ITALIAN: entello

Description

Average length: male head plus body 30 in (76.5 cm), tail 35 in (88.5 cm).
Average weight: male between 20 and 46 lb (9 and 20.9 kg); female between 13 and 40 lb (5.9 and 18 kg).

The silhouette is relatively slender but potbellied. The adult is gray, and the infant, until four or five months, brown. Face and ears are black, side-whiskers white or grayish. A fringe of hair forms a kind of volute behind the eyebrows. Hands and feet, also black, are long and slender with a short thumb and big toe. Some place it in a separate genus, *Semnopithecus*.

Family

Cercopithecidae.

Locomotion

While other species of the genus *Presbytis* are primarily arboreal, the hanuman langur spends 80 percent of its days on the ground. However, it is still very agile in the trees, moving on four limbs or by semibrachiation, and making horizontal leaps of 9.8–16 ft (3–5 m) and falls reaching 43 ft (13 m).

Diet

The highly developed salivary glands and the complex stomach of the hanuman langur are evidence of its foliverous diet. This does not prevent it from also eating fruits and flowers and various cultivated plants on occasion. It can cover several miles per day in search of food.

Predators

The tiger, the leopard, snakes, and dogs.

Longevity

In captivity, the hanuman langur lives approximately 25 years.

Distribution

Southern Tibet, Nepal, Sikkim, Kashmir, India, Bangladesh, Sri Lanka. The habitat of the hanuman langur ranges from sea level to 2.5 mi (4 km) high in the Himalayas, from tropical forests through jungles, temples, fields, and bazaars.

Status

Still widely distributed, the hanuman langur is endangered by the destruction of its habitat in favor of agriculture and forestry. In addition, while it is still sacred to the Hindu religion, its incessant pillaging of the crops of an ever-increasing human population is less and less tolerated.

Social Organization

The hanuman langur lives in unimale or multimale groups consisting of up to 125 members, with a ratio of two to six females per male. In the morning, the dominant male emits a cry calling together all the members, while at the same time keeping neighboring troops at a distance. He also decides in what trees the troop will look for food and sleep. In mixed troops, the relationships are peaceful, but there are also bands consisting of extra males that can attack them. After having chased away the resident males, they kill the infants so that all of the females will be in heat. Thus, in their new positions, they are certain to protect their own descendants and not those of their predecessors.

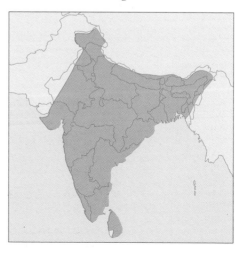

Hanuman Langur

Behavior

The activities of the hanuman langur occur early in the morning and late in the afternoon, with a siesta during the hottest hours required. Grooming occupies an important place in the social life. The hanuman langur spends up to five hours a day grooming. In Sri Lanka, it may occasionally do so in the company of another species, the toque macaque, with which it often plays, rests, and forages. Outside of these regular hours, its behavior varies according to the environment. Great variations occur, as the species occupies such diverse areas as desert edges, tropical forests, and alpine scrub. The environment influences the size of troops, the areas of distribution, the relationship between the sexes, dominance, sexual behavior, birth season, and even vocalizations.

Reproduction

The menstrual cycle is approximately 30 days, with the gestation period lasting from 196 to 210 days. A single infant is the rule. In the case of twins, the mother synchronizes nursing. All females participate in petting the newborn, while males do not play any part in their upbringing. The youngster is weaned at 10 to 12 months, during which time the mother does not ovulate. After that, it will not have any more individual contact with her. At three or four years, females reach sexual maturity, the males at four years, with their canine teeth not reaching maximum size until six or seven years.

History

Although it cannot be confirmed that this is the direct ancestor of the hanuman langur, fossil remains have been found near Athens, dating from 9.5 million years ago, of a species that strongly resembles it, only larger.

Fossilized remains of other members of this species' family date from the Oligocene and Pliocene in Africa and from the Pliocene in Asia and Europe.

A single infant is the rule among the hanuman langurs. And all of the females in the troop pamper the newborn.

Observation Site

India (Rajasthan): Ranthambhore National Park

Ranthambhore National Park, which covers over 108,680 acres (44,000 ha) of forest, is one of the most visited parks in India. It is located in the state of Rajasthan, in northwestern India. Its outline is characterized by a succession of undulating hills, which rise without transition above the enormous adjacent plain. It is in these valleys that most of the great mammals take refuge when the dryness affects the forest. The bottoms of many valleys are occupied by swamps, the largest, and the most spectacular, being artificial; they were, in fact, created several centuries ago by the maharajahs to attract game to what was at that time a hunting ground.

What, in the eyes of many visitors, constitutes the principal richness of Ranthambhore is the extraordinary mix of nature and human history; throughout the forest it is possible to admire the ruins of structures built by man: vestiges of hunting or pleasure lodges, ramparts, portals, burial places, and so on. The forest is, in addition, dominated by an imposing rocky outcrop, along which arises the ancient fortified city that gave its name to the national park. From high on the ramparts, the view of the forest is unique; inside the old walls, one has an indescribable feeling when faced with monuments that have been gradually conquered, by tenacious vegetation, and that seem to be waging one last, futile battle by desperately holding on to a lost past. A temple dedicated to the elephant god Ganesha is located within the walls of the old city, and pilgrimages are regularly made there by a crowd that is often as colorful as it is numerous.

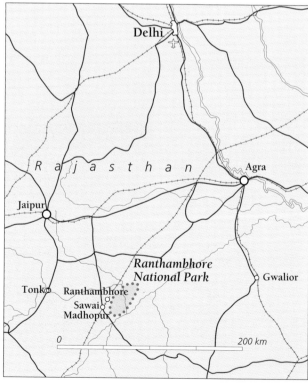

Fauna and Flora

The Ranthambhore forest is deciduous. During the dry season (our winter), the majority of the trees lose their leaves, which gives the site a rather desolate look but makes it easier to observe the fauna.

A wide variety of animals populate the Ranthambhore National Park, the tiger being obviously the most remarkable. The leopard is notable for its secretiveness, as is the sloth bear, which haunts the depths of the forest. *Chitals* and *sambars* (species of deer), nilgai antelopes, and wild boar constitute the principal observable species, to which can be added jackals, cyons (wild dogs), and, of course, monkeys, such as hanuman langurs and macaques. In the swamps it

The permanent swamps of the Ranthambhore National Park provide the food, drink, and coolness the animals need during the dry season, such as for this sambar *(Cervus unicolor)*.

Created in the lower valley to attract game, the swamps of Ranthambhore have exceptional biological richness.

is often possible to see impressive Indian crocodiles, as well as marine turtles of considerable size. These Indian crocodiles are also called marsh crocodiles or muggers, and are best observed during the dry season.

Observation

Widely distributed and often numerous, hanuman langurs can be seen throughout India, often in the immediate proximity of human installations. This is particularly true in certain religious temples where, along with the macaques, they seem to reign as uncontested masters. But the possibilities for observing hanuman langurs is particularly interesting in the Ranthambhore National Park. To see entire families of these agile primates evolve in the ancient fortified city is a spectacle that is quite unique. It is as if the hanuman langurs had become the new pathetic masters of a site long past its glory. Both curious and timid, during pilgrimages, they get very close to humans, as if they knew they were protected by the gods at these particular moments.

The Ranthambhore National Park is also an excellent place for observing hanuman

PRACTICAL INFORMATION

The best time of year to visit the park is during our winter, from November to February.

TRANSPORTATION

Flights to Delhi via Air India and other major carriers.

■ **BY CAR.** Ranthambhore National Park is located approximately 124 mi (200 km) southwest of the city of Agra. The main entrance is located approximately 9.3 mi (15 km) from the city of Sawai Madhopur, which can be easily reached, either by car or by train. Leaving Jaipur, an asphalt road makes it possible to reach Sawai Madhopur in approximately two hours.

■ **BY TRAIN.** Sawai Madhopur is located along the express rail line linking Delhi with Bombay, and all trains—or almost all—stop there. Several services are thus available daily.

■ **BY TAXI OR BUS.** From the city, it is easy to reach the park, either by taxi or public bus, which stops approximately 2.5 mi (2 km) from the entrance. It is also possible to rent bicycles, motorbikes, and jeeps in Sawai Madhopur. The majority of renters are located in the immediate vicinity of the train station.

ACCOMMODATIONS

Numerous possibilities for lodging, in all price ranges, exist in Sawai Madhopur. Within the park, the only lodging is the lodge called Jogi Mahal, formerly the maharajah's hunting lodge, admirably situated on the edge of the swamp near the main entrance to the park, This luxury establishment only has four rooms, however, which are often reserved long in advance.

Close to the park, another former maharajah's palace that was first transformed into a prison, Jhoomar Baori, has been made into a hotel that is admirably situated at the top of a hill, from where there is a magnificent view.

CLIMATE

A single word can be used to describe the climate that prevails in Ranthambhore most of the time—hot! Between November and February, however, the daytime temperatures fall on average to about 77–86°F (25–30°C) and the nights are bearable, while rain is rare. The monsoon arrives in July; it generally lasts until September-October.

TRAVEL CONDITIONS

Conditions for admission to the Ranthambhore National Park are rather strict and limiting: Two "safaris" are authorized each day, the first in the morning, the second in the afternoon. Outside of these periods, visitors' vehicles are not authorized to enter the park. There is an admission fee.

In the dry season (our winter), early mornings can be very cold when traveling by jeep open to the wind, so it is imperative to bring warm clothing.

The visit to the old fortified city, although it is included within the protected area of the park, is not subject to the same restrictions; access is authorized from sunrise to sunset, without requiring a guide, and it can be reached and explored on foot. However, you must have an admission ticket to the National Park to be able to reach the site.

India (Rajasthan): Ranthambhore National Park — Hanuman Langur

langurs in their natural environment and leading a life that is completely wild; numerous bands of monkeys haunt the forest of the park, confining themselves by choice to the rocky valleys within. Contrasts and similarities between the "city dwellers" and the "forest dwellers" can be easily observed here.

Hanouman, Warrior Monkey

By putting together, like a puzzle, the numerous tales of the adventures of the warrior monkey Hanouman, it is possible to recreate his history and his place in the thousand-year-old Sanskrit literature and that of the hanuman langur in the sacred Hindu tradition.

Hanouman, whose name means "broken jaw," was born of Vayu, god of the wind, and Punkikasthala, monkey-goddess capable of changing form. He grew up in her company, rarely seeing his father. But his mischievousness grew so that, using magic, the ancients momentarily made him lose his exceptional powers. He came under the tutelage of Surya, god of the Sun, whom he accompanied in his celestial travels. Gifted, well educated, and of exceptional eloquence, he was also an accomplished musician and a noble warrior, a hero who always followed the codes of war with respect.

Hanouman the warrior is gifted with an extraordinary power. Like his mother, he can change shape and make the earth tremble by beating his tail, but his sensitivity and his gentleness are also remarkable. In the Hindu tradition, what makes Hanouman as noble as he is strong is his ability, when his temperament vacillates between enthusiasm that makes him turn red to the deepest depression, to master himself and to fight against the worst adversities.

The bravura of Hanouman, the warrior monkey, explains perhaps the tolerance for hanuman langurs, despite their incursions into human populations.

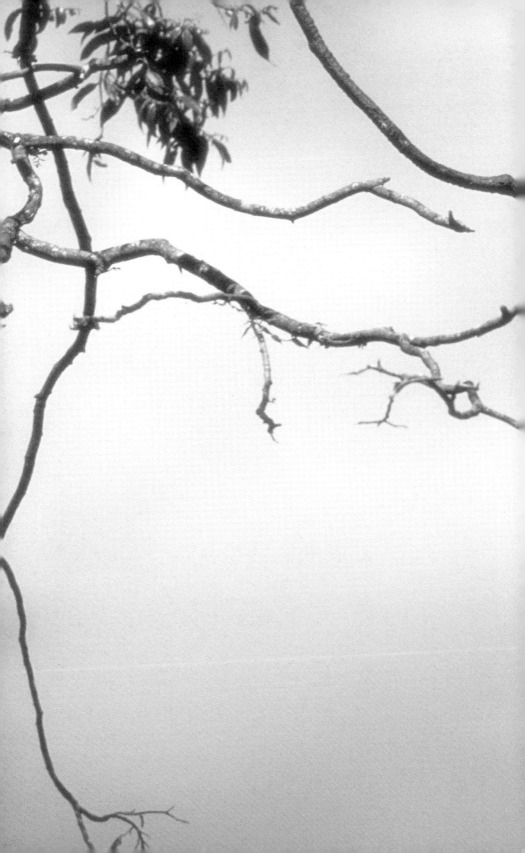

OBSERVING MONKEYS IN AMERICA

Black Howler
(Alouatta nigra)

FRENCH: hurleur de Guatemala
LOCAL NAME: mono aullador

Description

Average length: head plus body 22 in (57 cm), tail 24 in (60 cm).
Average weight: between 8.9 and 22 lb (4 and 10 kg).

The black howler has a sturdy body. Its face and the lower part of its prehensile tail are bare. The enlarging of the hyoid bone and the angle of the lower jaw make it possible for the howler to produce the impressive vocalization that gives it its name, and gives its throat a swollen look. The black howler is completely black.

Family

Cebidae.

Locomotion

Completely arboreal with a preference for higher or median levels of the primary forest, the howlers moves on four limbs. Its prehensile tail is so powerful that it can be used to hang from or to solidly grip a branch in the event of an accidental fall. This is fortunate, because, while they do not jump from branch to branch, these monkeys have the habit of diving from one forest level to the other.

Diet

All howlers feed on the barely ripe fruit they find on the small branches of emerging trees. Their large molars make it possible for them to digest leaves (40 percent of their diet), which they consume more than any other New World monkey.

Predators

Ocelots, jaguars, snakes, man.

Longevity

Howlers live on average 16 years, but can exceed 20 years.

Distribution

In its entirety, the genus of howler monkeys is the most widely distributed in the New World, ranging from southern Mexico as far as Brazil and Argentina. The black howler, itself, is limited to Mexico, Guatemala, and Belize.

Status

The howler population in Belize was severely reduced during the 1950s by yellow fever. Its situation is not critical, but it is still endangered and remains the rarest of its family.

Social Organization

While other species of howlers live in multimale troops of 3 to 45 individuals, even 65 for short periods, the black howler generally lives in small troops of from 4 to 10 monkeys, with a single male having a developed hyoid bone, the characteristic that determines hierarchy between males. The troop occupies a territory whose limits are redefined during movement by choruses of howls intended to be heard by neighboring troops.

Behavior

Howlers are up at sunrise, to define the territory of the

moment. During active hours, they can travel 1.9 mi (3 km) in the forest, and up to 3 mi (5 km) across lakes. Other vocalizations include the low chuckle of members coordinating the movements of the troop, the growling of disturbed adults, and the plaintive cry of females asking for help in finding a lost infant.

Feeding occupies up to 75 percent of their time, the rest being spent in social interaction.

Reproduction

Little is known of the reproductive habits of the howler. The smacking of the lips may be a prelude to mating, heat occurring every 13 to 24 days throughout the year. A single infant is born following a gestation period of 180 to 195 days, and it attaches itself to its mother's belly. Around four weeks, the tail becomes prehensile.

History

As with other New World monkeys, little is known of the history of the genus, other than that the earliest fossils date from the Miocene. With regard to the more distant past, archeological excavations made in the kitchen garbage of Mayan aristocrats have revealed the presence of some bones of howler monkeys (alongside more numerous human bones), indicating that on occasion they had a place at the table of kings.

The long and silky coat of the Guatemalan howler is a dark, shiny black, regardless of the age or sex of the individual.

Observation Site

Belize: Community Baboon Sanctuary of Bermudian Landing

The history of black howler monkeys in Belize is a perfect example of respectful and intelligently conceived tourism, which has effectively contributed to the protection of a wild species while, at the same time, benefitting local populations.

Credit for the success of the protection of howler monkeys in the Bermudian Landing area is due to one man, the American biologist Robert Horwich. Arriving at this small village at the beginning of the 1980s, he quickly realized that the howler monkeys that abounded in the neighboring forest could be effectively protected only if the village populations agreed to be actively involved in the process. He was convinced that there exist other means of protecting nature than in confining it in reserved areas protected against man, so he based his strategy in Bermudian Landing on the voluntary participation of the villagers in protecting the animals. He ended by convincing some of the inhabitants to participate in an experiment in land management that respects the forest environment. In order to do so, he took advantage of the traditional

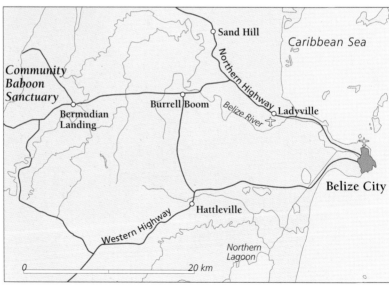

agricultural techniques practiced by the villagers, which cleared small forest areas where rice, beans, and then bananas were planted. After five years of cultivation, these clearings were abandoned. Alongside these agricultural parcels and the rivers, "corridors" of forest were maintained intact, making it possible for the monkeys to move and to quickly reuse the clearings left to lie fallow. At the end of a rotation period of ten years, the farmers once again took possession of clearings, which, in the interval, had recovered their fertility, and they once again cultivated them. The advantages of such a system are several: Erosion is stopped, harvests are better, and animals, which have since lost a good portion of their fear of man, are an important tourist attraction that brings in significant revenue for the villagers.

Less than 15 years after having been started, Robert Horwich's project has expanded over more than 12,350 acres (5,000 ha) and almost 18.6 mi (30 km) of the banks of the Belize River; approximately 900 howler monkeys live in the Community Baboon Sanctuary of Bermudian Landing, which has made it the largest population of howler monkeys actively protected in all of Central America. Such projects indicate the importance of weighing the needs of both animals and people during the design of conservation programs.

Fauna and Flora

The Bermudian Landing sanctuary is not, strictly speaking, a natural reserve in the classic sense of the term, but rather an agricultural area benefiting from management measures appropriate for nature conservation.

Birds are abundant, profiting from the varied environment that the alternation of cultivated expanses (open) with forest (closed) provides. Close to 100 species have been identified in the area. The most remarkable mammals that live in the primary or secondary forest in this area of Belize are the Virginia deer, Baird's tapir, and even the jaguar.

Caimans and marine turtles can occasionally be seen sunning themselves on the sand banks along the Belize River.

The Bermudian Landing Sanctuary has made it possible to effectively combine the protection of nature and the improvement of the economic resources of the inhabitants. Before any tour of the sites, a visit to the central office and information center is indispensible.

In addition to the howler monkey, the capuchin and the black-handed spider monkey are the two other species of primates that frequent these locations. Unlike the howler monkeys, these animals remain rather unobtrusive, and are, therefore, less often seen by visitors spending only a short time in Bermudian Landing.

Observation

Howler monkeys living close to man in the area covered by the protection project are clearly less timid than their cousins living in areas where open hunting is still causing damage. Several groups live in the immediate vicinity of the village of Bermudian Landing itself, and are used to visitors who have come to admire them.

It is, therefore, possible, with the help of a village guide, to observe the animals under excellent conditions, in their natural environment.

Living in an environment as dense as the lush primary forest of South America, the howler has developed an entire range of vocalizations to communicate with its fellow creatures.

The Howler Cries...

In the *Popl Vuh*, the mythological history of the Mayas, the monkey represents both the sacred, life after death, and the survival of an earlier world. According to legend, long ago, men made of wood were fishermen and the gods destroyed them in a flood, to replace them with men made of corn. Some men of wood, however, managed to escape the flood, and these are the monkeys.

The place of monkeys in the trees brings them close to heaven, while man remains attached to the earth. Two monkeys, in particular, live in Mayan territory, the spider monkey and the howler. The spider monkey, with its long limbs and its expressive face, is a joyful, sly creature, often appearing in Mayan art participating in erotic scenes, such as with its hand on a woman's breast. On the other hand, the howler, with its stocky build, and its serious air, is associated with duty and with death. Both worker and artist, it is seen turning ceramic vases or playing music.

The spider monkey laughs; the howler cries. Each is shown performing human activities, while man often appears engaging in the activities of apes. In classic Mayan art, on pottery as well as in painted murals, this ambivalence is evident. According to Rosa Raquel Romero de Barajas, the Mayas knew that in looking at monkeys, they were looking at themselves in a mirror, and this mirror reminds us that if the monkey can be another type of man, man is also a true species of monkey.

PRACTICAL INFORMATION

TRANSPORTATION

Belize City, the capital of Belize, can be reached only from another American country. The easiest connections are from Miami, Florida.

■ **BY CAR.** The visitor has various options for reaching the sanctuary: It is possible to rent a vehicle in Belize City (the capital of the country, 18.6 mi, 30 km away), hiring a taxi, or participating in a tour organized by one of the many agencies and operators in the city. The majority of organized tours to Belize include a visit to the Bermudian Landing sanctuary.

ACCOMMODATIONS

There are several opportunities for lodging in the immediate vicinity of the village, in small guesthouses run by the inhabitants; the level of comfort is relatively simple, but the welcome is warm. If not available, it is necessary to stay in Belize City, where the choice of establishments is clearly more extensive. It is entirely possible to make a round-trip visit to the sanctuary in a single day as the roads are in good condition in this country.

CLIMATE

The climate that prevails in the equatorial forest of Belize is markedly similar throughout the year. The temperatures, which reach 77–86°F (25–30°C) during the day, experience little seasonal variation. It can rain anytime throughout the year, but the rain is more frequent during the months of our summer. February is the coldest month, April being the warmest. During the rainy season, the fog can be dense in the forest in the morning after a rainy night.

TRAVEL CONDITIONS

Before beginning their visit to the sanctuary, visitors must go to the Administrative Center, located at the entrance to the village, where they can obtain an admission ticket—for a very reasonable fee—and be assigned a guide. The center has, in addition, a small exhibit on nature and the howler monkey protection project, which makes it possible to learn more about the site and the animals before the visit. Travel is on foot, over terrain that is sometimes rather difficult. Generally, it takes 20 minutes of hiking before encountering one of the groups that live in the immediate vicinity of the village, but the search for animals can occasionally take somewhat longer.

White-faced Capuchin
(Cebus capucinus)

GERMAN: Weisschulterkapuziner
FRENCH: capucin (or sapajou) à face blanche
COSTA RICAN: carablanca

Description

Average length: head plus body 17.7 in (450 mm), tail the same size.
Average weight: 2.4–7.3 lb (1.1–3.3 kg).

The capuchin has a stocky body, a round head, and limbs, especially arms, that are relatively short. As its name indicates, the whiteness of the hair on its face, but also on its shoulders and its chest, contrasts with its dark coat.

Family

Cebidae.

Longevity

Capuchins in captivity live up to 40 years of age.

Distribution

Present in Belize and Honduras as far as western Colombia and Ecuador, they are found at up to 6,888 ft (2,100 m) of altitude in the western Andes.

Status

With the exception of howlers, the subfamily of capuchins is the most widely distributed in the New World. White-faced capuchins are generally abundant.

Locomotion

The capuchin generally moves on all four limbs and is capable of real acrobatics. On the other hand, the tail is only partially prehensile, and serves, above all, to anchor the animal in the trees during rest, which is done in a seated position, on the side, or stretched out along a branch, arms and legs hanging at its sides.

Diet

According to a study conducted in Panama, the diet of the white-faced capuchin consists of 65 percent fruit, 15 percent leaves, 20 percent insects. This does not prevent it from coming down from the trees to pillage vegetable gardens and steal eggs. The local population compares the capuchin with a fox in a henhouse, which explains its lack of popularity among the villagers, and also its survival, despite the loss of its natural habitat.

Predators

Man, and, more rarely, raptors, wild cats, and snakes.

Social Organization

Capuchins generally form multimale troops, in which the females are nevertheless more numerous than the males. Troop movements are decided upon by an old male and take place in a precise order. Juveniles of both sexes go first, followed by adults, and finally, by pregnant females.

Their vocalizations include various chatterings, small sharp cries, or more piercing cries. The cries of a lost youngster alert the adults who go to look for it to bring it back to its mother.

Behavior

Capuchins spend the entire day foraging, except for taking a siesta during the hottest hours of the day. In

WHITE-FACED CAPUCHIN

coastal areas and in the absence of hunting or other forms of harassment, it is possible to find troops of 20 to 30 individuals that can be very bold toward humans. On the other hand, in the secondary forests of the interior, bands are smaller and run away from man in silence. Capuchins rub their feet and fur with urine to mark their territory, inside of which they travel on well-marked paths. They defend themselves with the aid of branches.

Their brain, with its highly developed convolutions, is also extremely large in relation to its body weight, two signs of great intelligence. Their flexible behavior allows them to exploit a variety of habitats, including wet and dry forests. In this, and by its great manual dexterity, the capuchin is similar to its Old World relatives. It is also for this reason that a neighboring species, the spider monkey, has been selected for experimental programs conducted in North America and in France; these monkeys are intended to provide assistance to tetraplegics, by performing up to 60 tasks on demand, such as turning lights on or off, taking a bottle out of the refrigerator, and so on.

Reproduction

The menstrual cycle is from 15 to 20 days with light bleeding, which is rare among New World monkeys. It is the females that solicit the attention of the males. Gestation lasts 180 days. Twins are rare. At the beginning, the infant attaches itself to its mother's belly, but later, it climbs on her back. When it begins to walk on its own, the mother holds it by the tail in order to prevent it from falling. The female is mature at approximately four years of age, the male at seven to eight years.

History

The first fossilized bones of a capuchin type date from the middle of the Miocene.

By the size and shape of its brain, as well as by its great manual dexterity, the capuchin resembles Old World monkeys.

Observation Site

Costa Rica: Santa Rosa National Park

The white-faced capuchin is one of the four species of primates present in Costa Rica—the others being the squirrel monkey, the black-handed spider monkey, and the howler monkey—and is perhaps the one that is the easiest to observe in various parks and reserves throughout the country. Rather eclectic in its choice of environment, the capuchin can, in fact, be found in the moist or dry forest, both on the plain and in average altitudes. The best places to observe these endearing primates in their natural environments remains, however, the dry forest, as can be found in Santa Rosa National Park.

This park is one of the largest in the country, extending over almost 98,800 acres (40,000 ha), covering a good portion of the Santa Elena peninsula, not far from the border with Guatemala. It is contiguous with Guanacaste National Park, thus constituting one of the principal conservation zones in all of Central America.

Santa Rosa was created at the beginning of the 1970s. It is meant to protect the natural riches of the area, but it also has a historical value, as it was there that a famous battle in Costa Rican history took place when a group of patriots victoriously defeated a troop of invaders commanded by the celebrated American plunderer William Walker in 1856. Today, Santa Rosa National Park protects one of the most important areas of low-altitude tropical forest in all of Central America. It is very popular in the country, both as a historic site as well as an important natural site. Many Costa Ricans make the historic pilgrimage to the house of La Casona, the last building still standing dating from the time of the Hacienda Santa Rosa, where the 1856 battle took place, and that today is a historical, cultural, and natural history museum.

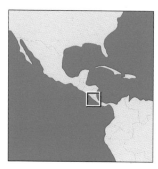

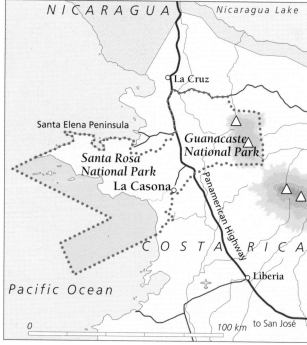

Fauna and Flora

The forest in Santa Rosa National Park extends as far as the edge of the beach. Many very different forest ecosystems are, in fact, protected inside the perimeter of the park: dry tropical forest, evergreen forest, swamp forest, and coastal forest. Parcels of wooded savannah, along with mangroves, complete the picture.

One of the most remarkable trees in the national park is the guanacaste, which has become the national tree of Costa Rica. It is a species typical of the forests of the Pacific coast of Central America, and can reach considerable size. The gumbo-limbo, another tropical American species, is also widely distributed throughout Santa Rosa. This tree is easily recognizable during the dry season, as the bark from the trunk has a tendency to dry out and fall in large pieces, leaving the deep red-orange trunk almost bare.

The avifauna of the park is represented by no fewer than 250 species, among which the parrots and the parakeets are often the most spectacular—and the noisiest. Santa Rosa is frequently visited by ornithologists, who can admire both species of forest bird and those typical of coastal and marine environments.

The park is also home to more than 100 species of mammals, among which bats are particularly well represented; more than 50 species, insectivores or frugivores, animate the nights with their flight. Deer are frequently seen (primarily the Virginia deer), the collared peccary, the coati, the nine-banded armadillo, the raccoon, the common opossum, sloths, and others. Primates are represented by white-faced capuchins, howler monkeys, and the black-handed spider monkeys.

The park also houses many species of lizards and snakes. The green iguana, common in pet stores in the United States, appears to be a different creature when viewed in its natural habitat. A six-foot male displaying high in a tree is a truly remarkable sight. The spectacled caiman lives in streams, ponds, and swamps, while the American

In the last few years, Costa Rica has acquired the reputation for being one of the best sites for eco-tourism in Central America. Many of its national parks combine the beauty of tropical forest with that of marine landscapes (here, in Guanacaste National Park).

Often more retiring than primates, reptiles abound in the tropical forest where 90 percent of them are unseen by visitors.

The colorful world of tropical birds is one of the major attractions of the natural sites in Central America. Shown is an orange-headed *Aratinga canicularis*, which makes its home in an ants' nest.

crocodile stays in the mangroves. Four species of marine turtles make intense use of the beaches of Santa Rosa to lay their eggs, the most frequent being the green and olive turtles.
It must be noted that in various places in the park, petroglyphs, which are assumed to date from the pre-Columbian era, decorate the rocks.

Observation

Observation conditions for fauna, and for capuchins, in particular, are very good in Santa Rosa National Park. An important network of trails, roads, and paths make it possible to cross the forest, and the guards generally know the best places to observe these monkeys and can escort visitors. The best season for observation is most assuredly the dry season, when the trees have lost most of their leaves, which makes it much easier to find the monkeys that often given themselves away by their rapid movement from branch to branch. An information center is open to the public in the historic La Casona house; it houses a very good educational exhibit on the animals and plants of Santa Rosa and the interactions between fauna, flora, environment, and climate.

A Common Pharmacy

In nature, the white-faced capuchin uses four kinds of medicinal plants: several species of *Citrus*, two lianas, *Clematis* (*barba de vieja* in Spanish), and *Piper* (anis) and a tree, the sloanea (*terciopela* in Spanish). But it does not use them as they are. The capuchin reserves a different preparation for each one and for each part that it uses (pulp or peel of the fruit, leaf, shoot, pod). It chews some of them, nibbles others, crushes or scrapes them, or even rolls them between its hands mixed with saliva. It then rubs its own fur or that of its companions vigorously with this preparation, using its hands, feet, or tail. It seems that many people in Central and South America use these same plants, according to the species, as disinfectants, fungicides, or mosquito repellents, in order to soothe insect bites, fever, rheumatism, colds, and flu, or to clean wounds. They do not, however, use the slonea, whose pods are irritating to man. Since insects do not attack beans in pods, however, it is assumed that the capuchins take advantage of properties we do not yet know how to use.

Perhaps the capuchin has an idea of the medicinal value of these plants, and rubs itself the way a cat rolls in catnip. It is odd that it is primarily in the rainy season, when the risk of insect bites, mycosis, and infections increases, that the monkeys engage in this activity.

PRACTICAL INFORMATION

During school vacations and weekends, Costa Ricans often visit the park in large numbers.

TRANSPORTATION

The airport in San José, the capital of the country, is linked daily via United Airlines from Los Angeles and Delta from Atlanta. The city of Liberia can be reached from San José by car, bus, or plane.

■ BY CAR. The entrance to the park is located some 18.6 mi (30 km) from Liberia, west of the Pan-American route leading to Nicaragua. A paved road exists in the park as far as the Administrative Center, where the historic La Casona house is located, as well as the official entry point and other service buildings. A road for all-terrain vehicles makes it possible to cross the park as far as the coast—approximately 6 mi (10 km). To reach the park, you can either rent a car in Liberia, or perhaps in San José, hire a taxi, arranging for the return trip, or use the frequent services of the bus that runs between Liberia and the Nicaraguan border.

■ BY ORGANIZED TOUR. It is also possible to participate in an organized tour; many agencies and tour operators can provide this type of trip, either departing from Liberia or from San José.

ACCOMMODATIONS

Camping is possible at the Administrative Center of the park near La Casona, as well as in several places along the beach. Campsites are safe but rustic, and generally have only rudimentary latrines with no showers. It is sometimes possible to obtain a place for the night at the Scientific Station located near the park's Administrative Center, depending on the number of scientists and students staying there. A small snack bar, also located near the Administrative Center, provides the sole possibility for dining in the National Park, so it is better to bring your own food.

CLIMATE

The climate in Costa Rica is characterized by two distinct seasons: dry from December to April, and humid the rest of the year. In the coastal area, however, it can rain throughout the year. The temperature stays around 77–86°F (25 to 30°C).

TRAVEL CONDITIONS

There is an admission fee to the park, in addition to a charge for each night spent within the limits of the protected area. While it is possible to travel around the park without being accompanied by a guide, it is often better to pay for the services of a guide, at least during the first trip, to improve your chances of being able to observe the animals, particularly the capuchins. It is possible to drive on the few roads open to traffic (a 4 × 4 is necessary) during the dry season; the rest of the year, travel in the park is possible only on foot or on horseback—horses can be rented at the Administrative Center. Travel at night is prohibited.

Red-backed Squirrel Monkey
(Saimiri oerstedii)

FRENCH: saïmiri à dos rouge
LOCAL: fraile, frailecito, barizo

Description

Average length: adult head plus body 12.6 in (32 cm), tail 16 in (41 cm).
Average weight: male between 19 and 40 oz (550 and 1,135 g); female between 12.8 and 26 oz (365 and 750 g); newborn 3.5 oz (100 g).

Squirrel monkeys are lithe, with a short, gray coat and nonprehensile tail. The red-backed squirrel monkey is distinguished from the common squirrel monkey by its redder coloration.

Family

Cebidae.

Locomotion

The squirrel monkey jumps very little but runs on all four limbs in the primary, secondary, or gallery forest, often near running water. While it prefers intermediate levels of the forest, it also exploits the canopy or even descends to the ground and cultivated fields. It can be very noisy, but also very quiet, and its movement is often noticeable only by the falling of leaves.

Diet

The major portion of its diet consists of berries, nuts, flowers, buds, seeds, leaves, and gums. Approximately 20 percent of its diet consists of insects, spiders, snails, and even tree frogs. On the ground, squirrel monkeys eat land crabs; higher up, they catch flies and butterflies.

Predators

Raptors represent a great danger for infants carried on the back. It is assumed, however, that the rather long gestation period for this species (165 days), which results in the birth of an already well-developed infant, serves to shorten the period of childhood, and thus the period of risk.

Longevity

Approximately 15 years, and up to 20 in captivity.

Distribution

The red-backed squirrel monkey is limited to Costa Rica and Panama, while the common squirrel monkey is found in the majority of rain forests in South America.

Status

The red-backed squirrel monkey is endangered; its close relative, the common squirrel monkey, is probably the least endangered of all the monkeys in the Americas, despite its exploitation as a pet and laboratory animal.

Social Organization

Squirrel monkeys form the largest troops of any of the New World monkeys, numbering 20 to 35 individuals in the eroded forests of Panama to more than 300 in the virgin forest of the Amazon basin. Territories are poorly defined, and no confrontation between troops has yet been noted. Mutual grooming has never been observed. The dominance of males is expressed by the exhibition of genital parts. During the day, troops disperse in bands of male adults, mothers with their young, and juveniles. During the mating season, males and females freely intermingle. Males store fat in their forequarters to give themselves a more imposing look, and ferocious confrontations can take place.

Pregnant females often form semi-independent subgroups within the main troop, and may establish their own dominance hierarchies. Adult males are excluded from these groups by the pregnant females. Pre-adult males may form separate units at this time as well.

Behavior

In relation to its size, the squirrel monkey has the largest brain of any monkey. As with many New World species, the human ear can easily discern its large repertoire of vocalizations. Babbling and screeching serve as warnings of contact or as alarms; cooing and raucous cries resound during the mating season; yapping indicates aggression. Squirrel monkeys are among the most vocal of the primates. Researchers have identified 26 separate calls.

Reproduction

Reproduction is strongly linked with climatic factors. The mating season is short, from September to November. Births are extremely synchronized from February to April, the rainiest period of the year; thus, youngsters begin to forage alone at the moment when arthropods, flowers, and fruits are abundant, as the mother stops feeding it very soon. At 11 weeks, the youngster is capable of capturing mobile prey, and, at 16 weeks, it is completely weaned. Males do not take care of youngsters.

Females reach maturity at two and one-half years, males at four.

History

It is possible that the genus of squirrel monkeys is the most ancient form of monkey in the New World. The natives of Guyana consider the squirrel monkey the archetypical monkey and call it the *monkey-monkey monkey*.

Common and widespread throughout most of tropical South America, squirrel monkeys are much more rare in Central America.

Observation Site

Costa Rica: Manuel Antonio National Park

While the squirrel monkey is one of the most widespread and common primates in the forest areas of the Amazon basin in South America, the same is not the case in Central America, where this species occupies only a limited territory of some 4,971 sq mi (8,000 sq km) in the low-altitude moist forests along the Pacific Coast, in Costa Rica, and in Panama, more than 373 mi (600 km) away from the closest South American population of squirrel monkeys. The two distinct subspecies that "share" Central America are in rapid decline; the last census, made in 1996, showed approximately 3,000 *Saimiri oerstedii oerstedii* and 500 to 1,000 *Saimiri oerstedii citrinellus*. In the past, huge numbers of squirrel monkeys were exported for use in medical research and the pet trade. This practice has been largely curtailed. The red-backed squirrel monkey is considered endangered by the IUCN. The destruction of their natural habitat constitutes the principal danger for these animals, whose future appears particularly bleak at present.

One of the best places to observe a small population of these endearing monkeys is in Manuel Antonio National Park, a small park of approximately 1,606 acres (650 ha)

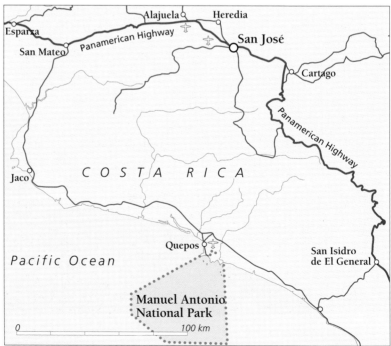

located along the Pacific Coast of Costa Rica, in the immediate vicinity of a coastal village bearing the same name. The latter is located about 7.5 mi (12 km) south of Quepos, one of the main coastal cities in northern Puntarenas province. The park, the smallest of the country's protected areas is, however, one of the most popular. Created only in 1972, just in time to avoid the destruction of the forest in order to construct hotels, this portion of the Costa Rican coast, strongly coveted for tourism, was subject to some very heavy financial speculation.

Fauna and Flora

The countryside in Manuel Antonio National Park is particularly agreeable. The forest extends as far as the wide, sandy beaches, and small rocky escarpments offer superb spots for viewing the sea and the islands located a short distance from the coast.

The forest in the park is a coastal type, which clearly differentiates it from the forest areas located more in the interior, generally at higher altitudes, and which can be found in Central America. Educational billboards located at the entrance to the park inform visitors about the most remarkable and most characteristic species of trees, bushes, shrubs, and plants inside. The *Hippomane mancinella*, a tropical tree locally called the "little apple tree," is among the most interesting in the coastal forest. Its fruits, actually resembling little apples, are poisonous, as is the liquid that oozes from the bark and the leaves. The poison that this liquid contains causes strong burning and itching sensations of the skin. In 1992 a particularly violent hurricane destroyed almost a quarter of the forest in the National Park; however, this so-called disaster was not actually one, since the natural forest that regrew in the devastated area is rich in plants that cannot develop in a mature forest; moreover, the areas of plant regrowth seemed to particularly suit squirrel monkeys. A tropical forest that has not suffered a hurricane or fire will have a closed, or nearly closed canopy. This means that little if any direct sunlight will reach the forest floor. Therefore, the types of plants that can grow there are severely limited, and only those that can toler-

Many of the protected natural sites in Costa Rica are fortunately not accessible by vehicles. Seen here, the Pacific coast jungle in Manuel Antonio.

ate shade will survive. Events such as fires or storms cause an influx of sunlight due to the downfall of large trees. Many sun-loving plant species then take root, fostering a great diversity in animal life as well. In this way, death and life, destruction and rebirth are an integral part of the great game of nature, in which often complex interactions remain largely unknown or misunderstood by man.

The islands located some distance from the coast are part of the national park. They constitute important reproduction sites for various species of marine birds, including the gannet and the pelican. A small coral reef extends some distance from the coast, where divers can admire a large number of tropical fish as well as a multitude of marine invertebrates and plants typical of coral environments. The diversity of fauna in Manuel Antonio National Park is surprising for a site that is so small in area. Many dozens of reptiles and amphibians are present there in large numbers. Some 350 species of bird, permanent residents or regular visitors, can also be observed in the forest and adjacent areas.

Mammals are also well represented; deer, pecaris, agoutis, armadillos, raccoons, and sloths are some of the species that may be encountered during a walk through the forest. The primates are represented by three species: squirrel monkeys, capuchins, and howler monkeys. The population of squirrel monkeys numbered 105 individuals at the time of the 1996 census.

Observation

The possibilities for observing fauna are very good in the park; the majority of animals have, in fact, become progressively accustomed to the intense human frequenting of the protected site and have lost some of their fear of humans, even adopting habits that are less nocturnal than those of their relatives living in areas where hunting remains an omnipresent danger.

Many visitors have the opportunity to see several species of mammals among those we have listed during a simple walk in the forest, and the monkeys are among the animals

Monkey Island

In 1978 the Pasteur Institute decided to become involved in research on malaria, and chose as its experimental model an animal resistant to the illness: the common squirrel monkey (*Saimiri sciureus*), close relative of the red-backed squirrel monkey. Two breeding areas were simultaneously established, one at the Institute in Cayenne, the other on a small island off the coast, Ilet-la-Mère, 138 acres (56 ha) in area. Successively inhabited by the Jesuits and convicts, the island was abandoned in 1945. When first occupied by man, a number of fruit trees were introduced (mango, guava, and mobins). Thanks to this abundance of fruit and the absence of predators, the island was a dream location for raising animals under healthy and natural conditions. Today, some of the monkeys live there in captivity, while some are free but still depend on the island's guardians for their food. Finally, some, that live there in complete freedom, are studied among others by researchers from the National Museum of Natural History. Bred monkeys are regularly freed and wild individuals captured, in order to avoid inbreeding. Those that are the subject of an experiment are first cared for, then released on the island. Tourism is authorized by prefectoral decree in a small portion of the island. Educational billboards present the biology and behavior of squirrel monkeys, their importance for research, and the efforts of the Institute to insure their well-being.

At the Pasteur Institute in Cayenne, French Guiana, the yellow squirrel monkey is studied for its natural resistance to malaria.

most frequently seen under good conditions, especially during the first hours of the day and at the end of the afternoon. For those who, despite everything, have not had the privilege of observing squirrel monkeys in their natural environment, a small rehabilitation station also exists inside the national park. The latter, known as Gaia Garden, is open to visitors. It is possible to observe the dozen squirrel monkeys that are kept there, and that are trying to return to the wild, a long and exacting operation.

It should be noted that the presence of tourist resorts near the park, and their popularity in Costa Rica, has as its inevitable consequence the heavy visiting of the site by tourists. Their number has increased to such an extent that the management of the park has been forced to take limitative measures to guarantee the calm necessary for the animals and to maintain sufficient quality for the visits. The park is presently closed on Mondays, and the maximum number of visitors authorized each day to enter the park is limited to 600; some sites within the park are even more strictly limited.

PRACTICAL INFORMATION

Be careful—some roads can be made very muddy, slippery, and difficult to travel by the rains, one even requiring the crossing of a relatively deep channel according to the time of the tides.

TRANSPORTATION
The airport at San José, capital of the country, is linked by daily flights from Los Angeles (United Airlines) and Atlanta (Delta).

■ BY BUS. Buses leave the capital, San José, twice a day during the dry season and at least once per day during the rainy season, in the direction of Manuel Antonio. The San José-Quepos Road can be very hard to navigate during the rainy season. Bus service from other cities also makes it possible to arrive in Quepos every day, and daily flights from San José also arrive there every day.

■ BY CAR. From Quepos, it is possible to reach Manuel Antonio either by hiring a taxi, by renting a car, or by using public buses. There is frequent service in both directions every day.

DISTANCES
San José to Quepos: 112 mi (180 km)
Quepos to Manuel Antonio: 6.2 mi (10 km)

ACCOMMODATIONS
There are several possibilities for lodging, adapted to all types of budgets, in Manuel Antonio and vicinity as this portion of the coast, in fact, attracts many tourists.

CLIMATE
The climate of the Pacific Coast of Costa Rica is tropical and humid, with temperatures wavering between 77 and 95°F (25 and 35°C) practically all year long, the dry season, however, being slightly less hot than the humid season. The temperature goes down only a bit at night.

Costa Rica experiences two distinct seasons: The rainy season extends from August to November, and the dry season from January to March. The rest of the year can be considered intermediary periods. Along the coast, however, these climate differences are often less clear, and it can rain anytime throughout the year, although most precipitation occurs during the rainy season.

TRAVEL CONDITIONS
Manuel Antonio National Park is accessible year-round, daily except Mondays, between sunrise and sunset. There is a reasonable admission fee, the ticket being good for the whole day. It is not necessary to be accompanied by a guide to walk through the park, but it is forbidden to leave the many open pedestrian trails in the forest. Travel is possible only on foot.

As we have mentioned, the large number of tourists visiting this small park called for restrictive measures as, for example, the prohibition against camping (including on the beach), the establishment of daily quotas for authorized visitors, and so on. Nevertheless, for those who are able to come to the park during midweek, the number of visitors present in the forest is tolerable, especially during the rainy season, during which the number of visitors considerably diminishes.

No infrastructure is available in the park, with the exception of an information center located near the main entrance.

MONKEYS, APES, AND MAN

Respectful Tourism

Tourism can, in the case of monkeys, be the best or the worst thing. The creation of structures making it possible for men and monkeys to come into contact with each other almost systematically runs up against a certain opposition on the part of scientists and pure and hard-line "conservationists," who emphasize the potential risks that visiting the animals can entail for their safety. Well aware of the potential dangers represented by tourism for certain species of primates, we consider that such activities, when they are well organized, limited, and strictly controlled, are definitely more positive than they are negative. Trips to view the mountain gorillas in central Africa, which were strongly opposed by the well-known researcher, Dian Fossey, for example, who relentlessly defended the cause of these animals, clearly demonstrates that when tourism is intelligently conceived, it can greatly contribute to the protection of the monkeys concerned. This benefit arises chiefly from the greater appreciation felt for wild creatures by those who have observed them under near-natural conditions. Also, well-managed programs can assist conservation efforts by encouraging local people to see that it is to their benefit to protect wildlife. It is unrealistic to expect that commercially valuable land and animals can be protected without offering some benefit to the people who live in the area and who may have traditionally used natural resources for their survival. If such people are given the opportunity to earn a living within the framework of conservation, the animals so protected take on a new value and hence there is a reason, in the view of those living nearby, to protect them. Provisions should be made for sustainable use of the forest and its animals by those accustomed to living in this way, and who may not have other means of meeting their daily needs.

On the other hand, when tourism is excessive and uncontrolled, it can prove to be disastrous. Constructing a permanent route for tourists to arrive en masse in such a forest environment in order to see the animals can, for example, cut the traditional migratory paths of arboreal monkeys, depriving them of food for part of the year, and the presence of too great a number of humans in the environments of particularly timid species constitutes a direct menace for their survival. Thus, we have been forced to list here primate observation sites that cause a minimum of negative secondary effects; whether it is a question of a species that easily adapts to the presence of humans such as howler monkeys in Belize, for example, or of organizing visits that are strictly controlled, such as those to see mountain gorillas in Uganda. It is, however, very important that all visitors follow a code of ethics and behave respectfully in the presence of primates or that they come into contact with in their own environment. Contacts that are too close are to be avoided at all cost. We must remember that primates are capable of contracting many of the illnesses that affect the human species, beginning with the common cold; moreover, monkeys that are too used to close contact with humans often develop behaviors that are particularly detestable, such as stealing, attacking people, pillaging of tents in camping areas, and other habits. In the national parks of southern Africa, each animal guilty of reprehensible acts (this concerns primarily baboons) is systematically destroyed by the guards. Tourists that accustom animals to receiving food thus irrevocably condemn them to death! In addition, giving any kind of food to monkeys must be avoided, as they are perfectly capable of meeting their own needs themselves. This is particularly true in such sites as Gibraltar (Barbary macaques) or Jigokudania (Japanese macaques). Offering chocolate, cookies, french fries, or ice cream to animals, in the long run proves to be as fatal to them as gunshots, traps, or poison.

Finally, the conscientious and respectful visitor abstains from provoking the monkeys in any way, and from behaving in ways that are capable of stressing, frightening, or angering them. Better yet, do not hesitate to clearly demonstrate—with courtesy—to other visitors, and even to guides, that some of their habits are not appreciated—provoking a vocal dual with the dominant male of a troop of monkeys, for example. If this type of practice pleases certain tourists yearning for "strong" impressions à la King Kong, they are often seriously disturbing to the animals, with the negative consequences that can result, such as increased aggressiveness within the troop, repeated escapes, stress, influence on gestation, and so on. Seeing primates under good conditions and in a short amount of time must be considered a real privilege, which must be reserved for a certain "elite" that is conscious and respectful of the animals. These animals offer, in exchange, memorable moments and unforgettable memories for those who make the effort to come as guests, and not as invaders.

A Death Sentence?

More than half of the species of primates are today represented on the Red List of Endangered Species established by the World Conservation Union. An unhappy balance sheet—primates, which include the closest evolved relatives of the human species, have, in fact, suffered from the activities of humans throughout the world; while some species have been able to adapt to the changes brought by humans to their environment, others—often the most specialized—are today approaching the threshold of extinction.

The destruction of natural habitats, principally the forest, currently constitutes the principal danger to the survival of numerous species of primates in all the tropical and subtropical regions of the world. A large majority of these animals are, in fact, inhabitants of the forest belt that extends around the world on both sides of the equator; however, this is one of the natural environments that has been subject to the highest degree of destruction on the part of man for more than a century. Uncontrolled deforestation affects the lemurs of Madagascar as well as the orangutans of Borneo, the chimpanzees of western Africa, and the uakaris of the Amazon.

The selective exploitation of wood in the tropical forest makes it possible, however, in many cases, for primates to adapt to the changes in their environment that this entails, and thus to survive. Studies have even showed that certain species benefit in one way or another from changes in the flora in the areas of secondary forest that develop following selective forest exploitation operations. The same is not true when large expanses of forest are cut down. The destruction of the forest to make way for agriculture or industrial plantations of coconut palms, palms, heveas, and others gives primates little chance, with only a few exceptions.

Forest fires, caused in a majority of cases by man, can also prove catastrophic for primates. The terrible fires in Indonesia in 1997, for example, seriously affected the populations of orangutans in the areas concerned. While we still do not know how many of these great anthropomorphic apes were lost, it is known that dozens among them were killed by the inhabitants when, deprived of habitat and food, they penetrated into plantations or farms in desperation. At the end of 1997, rehabilitation centers existing on the island of Borneo were receiving baby orangutans whose mothers had just been killed—at an average rate of one a week!

Hunting continues to claim a large number of victims among primates. In many areas of the world, primarily in South America and Western Africa, monkey meat is appreciated by the inhabitants, and these animals are intensely hunted to supply the trade in "meat of the bush." These practices, while not reprehensible in and of themselves, go beyond the level of recovery for most of the species involved. Due to demographic growth, the improvement in means of communication and the techniques for preserving meat, an activity that had remained an art until a short time ago, has today taken on

Building site of a new route across the tropical forest in French Guiana. Once completed, it will be the means of penetration for hunters, farmers, and others, jeopardizing an important source of oxygen for the planet and the habitat of a large number of primates.

the look of a real industry in several tropical areas. Many primates are also hunted to supply the trade in pets; in this case, very young animals, not weaned, are involved, but in order to capture them, it is necessary to kill the mother, and often other individuals in the troop that desperately try to oppose the hunters. At the same time, scientific laboratories, regardless of the scientific or commercial sector, are still and always responsible for the death of thousands of primates each year. Of course, the situation has improved considerably when compared to what was happening 10 or 20 years ago; main laboratories get their supplies from commercial breeders, which does not make their activities any more tolerable from the point of view of the primates concerned, but the fact still remains that certain species reproduce in captivity only with difficulty, and their breeding is long and costly. The capturing of individuals in nature remains a commercially profitable option. The same is true with regard to a certain category of zoos and private collections, which do not hesitate to order the capture of primates in nature in order to be supplied. "Fashion" can also have perverse effects—baby orangutans, for example, are very fashionable with a certain class of residents of Taiwan, who keep them as pets, which has given rise to significant traffic representing a direct threat to the survival of this species in Borneo.

In many regions where forests are shrinking, many primates are victims of conflicts with man. Deprived of sufficient food, or simply attracted by the fruits, vegetables, or cereals obligingly planted by farmers, monkeys are often forced to loot harvests, which results in their being mercilessly hunted. Too large a number of primates have had their numbers decline in an alarming way. When these numbers fall below a certain safety threshold, the survival of the population can be directly threatened by an epidemic, a natural disaster, such as a volcanic eruption, one caused by man, such as fires, or by political instability in the area where the primates live. The case of the mountain gorillas of the Virunga Mountain chain, in central Africa, is a striking example: Only a few years of merciless war between ethnic groups was enough to reduce to almost nothing decades of effort on behalf of these primates. If the troubles in the area continue in the years to come, the future of the gorillas there may become truly problematic. When the number of individuals in a population or a species is reduced to only a few dozen animals, the genetic impoverishment that results irrevocably condemns them to death and extinction.

Significant efforts have certainly been made for the purpose of protecting the primates of the world and of assuring their survival. The majority of species are today protected by international conventions (such as CITES), and the volume of international commerce involving these animals has thus been considerably reduced, but the same does not hold true, alas, for internal commerce in certain areas. Protected areas have been created, sheltering large forest spaces in tropical regions, and active campaigns of sensitization have changed the perception that inhabitants have of primates in various countries.

Rehabilitation centers constitute a last-chance operation for several species of primate in Africa, Asia, and tropical America. Rehabilitating monkeys to return them to life in the forest, in order to reinforce wild populations or repopulate sites where the species in question have disappeared, has proved to be an operation that is particularly

What makes primates attractive can also be their downfall. Seen here, a juvenile orangutan, victim of international trade.

arduous, long, and costly. It should, in no case, take priority over conservation efforts on behalf of primates in places where they still survive in nature, in spite of some success achieved by this method, specifically in the saving of several species of tamarins in the Atlantic forest of Brazil.

Greater efforts will certainly have to be made in the years to come if the survival of all species of primates is to be assured. A challenge, certainly, but not a utopia.

The scale of the problem is such that drastic, innovative steps are required. In many areas, enforcement of laws is difficult if not impossible due to the lack of funding and trained personnel, the persistence of overwhelming events such as war and revolution, the nature of the landscape, and the beliefs of local people. If the need and desire for primates as a food source cannot be curtailed by law, commercial rearing of the animals may be a viable alternative. While not without problems, captive rearing of valuable animals often takes the pressure off wild populations, as it is easier and cheaper than hunting in the long run. The plight of the American alligator is a classic example of this concept. Hunted nearly to extinction, captive rearing has now totally eliminated poaching, as it is no longer profitable. Raising monkeys for food would repel the well-intentioned but often poorly informed "animal rights" people, but the magnitude of the threats facing many species makes such considerations irrelevant. The fate of the species, not the individual, must take priority. It is unrealistic and elitist for wealthy urban dwellers to imagine that hungry people with few means of changing their circumstances are going to view an edible creature as worthy of protection in and of itself. Alternative means of meeting daily needs must be provided if large populations of primates and many other creatures are to be sustained.

Helping Primates

Why not combine the useful with the agreeable in the service of monkeys? More and more often, projects for the conservation of wild fauna call for volunteers motivated to contribute to their activities. This formula is advantageous for both parties; for the projects, it offers enthusiastic collaborators that make it possible to efficiently promote the protection of the species in question, while for the volunteers, it offers the possibility of having interesting and memorable experiences in the service of nature conservation.

In the specific case of monkeys, the possibilities of participating as a volunteer in conservation projects are rather limited. The International Ecovolunteers Network, however, offers many opportunities, in Africa and in Asia:

• In Thailand, volunteers can participate in the activities of the gibbon rehabilitation center established on the famous island of Phuket, in the Khao Phra Thaew National Park. The purpose of the project is to recover gibbons held by inhabitants in order to return them to life in the wild in the National Park forest. Volunteers participate in the daily care provided to the residents of the center, as well as in the monitoring of individuals returned to the wild.

• In India, volunteers can take part in the study and protection of macaques in the state of Tamil Nadu. The project involves a comparative study of groups of these primates living in the forest and others living close to human dwellings, in order to determine the macaques' capacities for adaptation to changes in their environment.

• Finally, in Sierra Leone, volunteers can participate in the activities of the chimpanzee rehabilitation center created several years ago by a local organization. The activities include the daily care of the residents of the center, the rehabituation of animals to natural conditions, and the monitoring of released individuals.

The International Network of Ecovolunteers is represented:
– in France by the association "À Pas de Loup," 48, av. Félix-Faure, 75015 Paris. Tel./Fax: 01 47 53 02 03.
– in Belgium and in Luxembourg by the association "Tierra," Heidebergstraat, 226, 3010 Leuven. Tel./Fax: (016) 25 56 16.
– in Switzerland by "Acatour," Bahnhofstrasse, 28, 6301 Zug. Tel.: (041) 729 1420 Fax: (041) 729 1421.

APPENDICES

Equipment

In the majority of cases, the observation of monkeys in their natural environment takes place under rather basic conditions; terrain, vegetation, climate, and other factors can make the trip rather uncomfortable and reduce the possibilities of observations.

Good hiking equipment is a general rule, including shoes adapted to the conditions of the terrain, such as hiking shoes in the mountains, lighter hiking shoes in the savannah and in the tropical forest, and clothing adapted to the climate conditions. In the mountains, do not forget that it can get very cold, even in Africa or in Asia, and it generally rains a great deal; therefore, you must have the appropriate equipment. Wherever you go, clothing in neutral colors is preferable, in order to be as unobtrusive as possible; it is foolish to transform yourself into a commando with military-style clothing, knives hanging from the belt, and so on, which are generally not appreciated by the local authorities in many tropical countries.

A small backpack for carrying items you may need during the day is always practical. Binoculars are indispensable if you wish to be able to observe the monkeys under the best conditions possible, and to get the most out of your visit.

Photographing Primates

As is often the case when it is a question of photographing wild animals, telephoto lenses are indispensable in order to take good shots of primates living free. The photographs that can be found in many books or magazines, each more surprising or remarkable than the next, can give the impression that it is easy to photographs primates, but nothing could be further from the truth.

With the exception of several very specific situations, such as wild gorillas or chimpanzees accustomed to the presence of visitors that allow themselves to be approached without fear, or vervets or baboons used to pillaging campgrounds and that do not hesitate to be approached themselves, and so on, primates are generally rather timid and do not allow themselves to be approached at the distance required to take closeups or portraits. You must show patience, in order to allow the animals time to get used to the presence of an observer, and to be able to photograph the natural behavior of calm and relaxed animals.

Something to protect your equipment against the rain, the dust, and the humidity, all factors that can rapidly deteriorate housings and lenses, is indispensable. In the tropical forest, bring high-speed film, as the foliage is often very dense and strongly limits the quantity of light filtering to the ground.

Organizations

- World Wildlife Fund (WWF):
  1250 24 Street, NW
  Washington, DC 20037

In France	WWF - France
151, boulevard de la Reine
78000 VERSAILLES
Tel.: (33) 01 39 24 24 24
In Belgium	WWF - Belgium
Chaussée de Waterloo, 608
1050 BRUSSELS
Tel.: (32) 2 340 09 99
In Switzerland	WWF - Switzerland
Postfach
8010 ZURICH
Tel.: (41) 1 297 21 21

- World Conservation Union (UICN)

The UICN has groups of specialists who are concerned with various categories of wild animals or plants. This is specifically the case with primates; various experts specialize in the study and the protection of primates in Africa, Asia, tropical America, and elsewhere.

- UICN—The World Conservation Union
  28, rue Mauvernay
  CH-1196 GLAND
  Switzerland

- International Primate Protection League (IPPL)
  IPPL
  P. O. Box 766
  Summerville, SC 29484

Glossary

Allogrooming: Mutual grooming between individuals. Often directed at hard-to-reach areas, it assists in skin care by removing dirt and parasites. Such grooming also cements relationships, thereby serving an important social function.
Angiosperm: Flowering plant. While angiosperms are dominant today, they arrived late on the evolutionary stage.
Appeasement: An action that inhibits attack, such as the sexual presentation posture of female primates.
Canopy: Highest part of dominant trees in a forest that, like a ceiling, prevents light from reaching the ground.
Catarrhini: Old World monkeys. Their nostrils open toward the bottom.
Cuspids: Points of molars and premolars.
Ecological niche: Role and place of an organism in the functioning of the ecosystem.
Emerging tree: Tree surpassing those that surround it in the forest.
Estrus: State of receptivity in the female.
Facial mimic: An animal with facial expressions that serve to communicate a particular condition.
Folivore: Just as a frugivore eats fruit, the folivore consumes leaves.
Frugivore: Fruit eater.
Gallery forest: Forest that extends along both sides of a river, forming a roof over the water. This is a type of forest much favored by primates, as food is often abundant there.
Hyoid bone: Horseshoe-shaped bone in the neck near the base of the tongue.
Ischial callosities: Hard skin on the thighs serving as natural cushions in the seated position.
Olfactory bulb: Organ for the reception of olfactory stimuli.
Opposability: Ability to put the end of the thumb and that of the index finger face to face, making it possible to seize objects with dexterity.
Palmigrade: Form of locomotion in which the palm of the hand is placed directly on the ground.
Perineum: Part of the body between the anus and the genitals.
Phylogenetic: Relating to the genetic history of a species during its development.
Platyrrhini: New World monkeys. Their nostrils open toward the side, sometimes giving a flattened appearance to the nose.
Polyandry: Rare social structure in which there are several males for a single female.
Prehensile: Able to grasp an object because of the flexion of the hand, the foot, or the tail.
Rhinarium: Hairless, sensitive, and moist skin that, with the exception of tarsiers, surrounds the nostrils in prosimians.
Sexual dichromatism: Difference in color between the sexes of the same species.
Sexual dimorphism: Morphological difference, most often on the level of size, between the sexes of the same species.
Species: Basic unit of classification. It is generally said that if two animals can produce a fertile offspring, they are from the same species.
Stereoscopy: Three-dimensional vision that is due to the superimposition of two images of the same subject taken simultaneously by two parallel lenses (in this case the eyes).
Taxonomy: Theoretical study of the classification of organisms.

To Learn More

Books

General Works

A.F. Coimbra-Filho, R.A. Mittermeier, *Ecology and Behavior of Neotropical Primates*, Academia Brasileira de Ciencias, Rio de Janeiro, 1981.
R. Corbey, B. Theunissen, *Ape, Man, Apeman: Changing Views since 1600*, Proceedings of the Symposium held in Leiden, the Netherlands, 1993.
F. DeWaal, *Chimpanzee Politics: Power and Sex Among Apes*, Johns Hopkins, 1989.
R. G. Gelman, *Monkeys and Apes of the World*, Watts, 1990.
C.P. Groves, *Macaques: Studies in Ecology, Behavior and Evolution*, D. G. Lindburg, 1980.
J. McCrone: *The Ape that Spoke*, Avon, 1992.
W. C. McGrew, et al, eds., *Great Ape Societies*, Cambridge University Press, 1996.
J. Milton, *Apes: They're Like Us*, Random House, 1997.
K. Milton, *The Foraging Strategy of Howler Monkeys*, Columbia University Press, 1980.
J. R. Napier and P. H. Napier, *The Natural History of the Primates*, MIT Press, 1994.
P. Oberlé, *Madagascar, un sanctuaire de la nature*, Diffusion Lechevalier SARL, 1981.
D. Premack and A. Premack, *The Mind of an Ape*, Norton, 1984.

Appendices

B.B. Smuts, D.L. Cheney, R.M. Seyfarth, R.W. Wrangham and T.T. Struhstaker, *Primate Societies*, Chicago University Press, 1986.

S. Savage-Rumbaugh, et al, *Apes, Language, and the Human Mind*, Oxford University Press, 1998.

The Great Apes

M. Leach, *Great Apes*, Sterling (U.K.), 1998.

S. Montgomery, *Walking with the Great Apes: Jane Goodall, Dian Fossey, Birute Galdikas*, HM, 1992.

M. Nichols, *The Great Apes: Between Two Worlds*, National Geographic, 1994.

H. Preuschoft, D. J. Chivers, W. Brockelman and N. Creel, *The Lesser Apes*, Edinburgh University Press, 1990.

Primate Index

Alouatta nigra, 98–99
Antipredator behavior, 16
Auditory communication, 23

Barbary Macaque, 38–39
Birth, 27
Black Howler, 98–99
Books about primates, 125–126
Brachiation, 20–21
Brain, 11

Callitrichidae, 14
Catarrhinian monkeys, 14
Cebidae, 14
Cebus capucinus, 104–105
Celebes Black Macaque, 79–80
Cercopithecidae, 14
Cercopithecines, 14
Cercopithecoids, 14
Characteristics, 10–12
Cheirogeleidae, 57
Chimpanzee, 24, 43–44
CITES, 120
Classification, 13–15
Colobins, 14
Colobus polykomos, 30–31
Common Gibbon, 68–69
Communication, 23–25
Conservation projects, 121

Daubentoniidae, 58
Deforestation, 119
Diet, 16–19

Distribution, 13–15
Dominance, 25–26

Evolution, 10

Family group, 22–23
Forest fires, 119
Forest habitat, 16

Gelada, 49–50
Gestation period, 27
Glossary, 125
Gorilla gorilla beringei, 34–35

Hand, 10
Hanuman Langur, 90–91
Hapalemur griseus, 64–65
Heat, 26
Hierarchy, 25–26
Hominians, 15
Hominids, 15
Hominoids, 14
Hunting, 119–120
Hylobates lars, 68–69
Hylobatidae, 15

Indri, 59–60
Indriidae, 57
Indri indri, 59–60

Japanese Macaque, 74–75

Lemur catta, 63–64
Lemurians, 21
Lemuridae, 57
Lemuriforms, 14
Lemurs, 57–65
Lesser Bamboo Lemur, 64–65
Locomotion, 20–22

Macaca
  fuscata, 74–75
  nigra, 79–80
  sylvanus, 38–39
Mating, 26–27
Megaladapidae, 57–58
Meissner corpuscles, 10
Mountain Gorilla, 34–35
Multimale troops, 22

New World monkeys, 14

Old World monkeys, 14
Olfactory communication, 24
Orangutan, 84–85
Organizations, 124

Appendices

Paninians, 15
Pan troglodytes, 24, 43–44
Platyrrhinian monkeys, 14
Pongids, 15
Pongo pygmaeus, 84–85
Presbytis entellus, 90–91
Primates:
   antipredator behavior in, 16
   characteristics of, 10–12
   classification and distribution of, 13–15
   communication of, 23–25
   diet of, 16–19
   dominance and hierarchy in, 25–26
   evolution of, 10
   life in forest and savannah, 16
   locomotion of, 20–22
   photographing, 124
   reproduction of, 26–27
   social life, 22–27
   troop composition, 22–23
Propithecus verreauxi verreauxi, 61–62
Prosimians, 13–14
Purgatorius ceratops, 10

Quadruped locomotion, 20

Red-backed Squirrel Monkey, 110–111
Rehabilitation centers, 120–121
Reproduction, 26–27
Ring-tailed Lemur, 63–64

Salmiri oerstedii, 110–111
Savannah, 16
Sexual maturity, 27
Simians, 14–15
Social life, 22–27
Specialization, 10
Stereoscopic vision, 11

Tactile communication, 23–24
Tarsiforms, 14
Theropithecus gelada, 49–50
Tourism, 118–119
Troop composition, 22–23

Unimale troops, 22

Verreaux's Sifaka, 61–62
Vertical locomotion, 20–21

Western Black-and-White Colobus, 30–31
White-faced Capuchin, 104–105
White-handed Gibbon, 68–69

Geographical Index

Belize, 100–103
Berenty Reserve, Madagascar, 58
Bermudian Landing, Belize, 100–103
Boabeng-Fiema Sanctuary, Ghana, 32–33
Bohorok Rehabilitation Center, Indonesia, 89
Bwindi Forest, Uganda, 36–37
Community Baboon Sanctuary of Bermudian Landing, Belize, 100–103
Congo, 36–37
Costa Rica:
   Manuel Antonio National Park, 112–115
   Santa Rosa National Park, 106–109
Ethiopia, 51–53
Ghana, 32–33
Gibraltar, 40–42
Gombe Stream National Park, Tanzania, 45–48
Gunung Leuser National Park, Indonesia, 86–89
Honshu, Japan, 76–78
India, 92–95
Indonesia:
   Bohorok Rehabilitation Center, 89
   Gunung Leuser National Park, 86–89
   Tangkoko Duasudara Natural Reserve, 81–83
Japan, 76–78
Jigokudani Onsen, Japan, 76–78
Ketambe Scientific Research Center, Indonesia, 89
Khao Phra Thaew National Park, Thailand, 70–73
Khao Yai National Park, Thailand, 70–73
Madagascar, 56–58
Mahale Mountains, Tanzania, 48
Manuel Antonio National Park, Costa Rica, 112–115
Monkey Hill, Gibraltar, 40–42
Perinet Reserve, Madagascar, 58
Rajasthan, India, 92–95
Ranthambhore National Park, India, 92–95
Rwanda, 36–37
Santa Rosa National Park, Costa Rica, 106–109
Simian Mountains National Park, Ethiopia, 51–53
Sulawesi, Indonesia, 81–83
Sumatra, Indonesia, 86–89
Tangkoko Duasudara Natural Reserve, Indonesia, 81–83
Tanzania, 45–48
Thailand, 70–73
Uganda, 36–37
Virungas Chain: Congo, Rwanda, Ghana, 36–37

Acknowledgments

Our warmest thanks to Nicole Vallée and Bruno Toussaint for their kind and patient editing, as well as Professor Pujol of the ethnobiological library of the National Museum of Natural History, Bernadette Gilbertas, and Dr. Maria Santini for their documentary advice.

This work was accomplished with the collaboration of
Anne Cauquetoux

© Copyright 2000 by Editions Nathan, Paris, France.

English language version © Copyright 2000 by Barron's Educational Series, Inc.

Title of the original French edition:
Cap sur les Singes et les lémuriens.

All rights reserved. No part of this book may be reproduced in any form, by photostat, microfilm, xerography, or any other means, or incorporated into any information retrieval system, electronic or mechanical, without the written permission of the copyright owner.

All inquiries should be addressed to:
Barron's Educational Series, Inc.
250 Wireless Boulevard
Hauppauge, NY 11788
http://www.barronseduc.com

International Standard Book
No. 0-7641-1163-9

Library of Congress Catalog Card
No. 99-73358

Printed in Italy by
Arti Grafiche E. Gajani S.r.l.
Rozzano (Milano)

Iconographic Research
Rémy Marion

Photo Credits

C. BALCAEN: p. 91, 93t, 93b
T. BANGUIN (WWF International BIOS): p. 86
I. BARTUSSEK: p. 31, 122–123
S. BONNEAU: p. 99, 105, 107, 108t, 108b, 111, 113
C. ET M. DENIS-HUOT: p 18b, 25t
P. DE WILDE: p. 47
A. ENDEWELT (Objectif nature): p. 88
J.-P. FERRERO (PHO.N.E.): cover, p. 6, 19, 25, 66–67
J.-P. FERRERO/J.-M. LABAT (PHO.N.E.): p. 43, 46b
O. GRUNEWALD: p. 8–9, 13, 18t, 96–97, 102, 116-117, 119
M. GUNTHER (BIOS): p. 34, 37, 46t, 50, 52, 57b
H. HAUTALA: p. 75t, 78t
M. HEUCLIN (BIOS): p. 65
J.-M. LABAT/C. JARDEL (PHO.N.E.): p. 57t, 59
J.-F. LAGROT: p. 22, 54, 60, 61, 62, 63, 82t, 82b, 87, 114
S. LAHUZIER, p. 77b
R. MARION (PÔLE D'IMAGE): p. 42, 77, 88b
MINDEN PICTURES (STOCK IMAGE): p. 27, 120
NATIONAL GEOGRAPHIC (STOCK IMAGE): p. 28
G. PLANCHENAULT (SCIENCE ET NATURE): p. 71, 72
C. RUOSO (BIOS): p. 39, 41, 79
R. SEITRE (BIOS): p. 70, 95
R. VALTER (PHO.N.E.): p. 19b
A. VISAGE (PHO.N.E.): p. 84
J. ET P. WEGNER (JACANA): p. 69
P. WEIMANN (BIOS): p. 101